Springer Texts in Statistics

Advisors:
George Casella Stephen Fienberg Ingram Olkin

Springer
New York
Berlin
Heidelberg
Hong Kong
London
Milan
Paris
Tokyo

Springer Texts in Statistics

Alfred: Elements of Statistics for the Life and Social Sciences
Berger: An Introduction to Probability and Stochastic Processes
Bilodeau and Brenner: Theory of Multivariate Statistics
Blom: Probability and Statistics: Theory and Applications
Brockwell and Davis: Introduction to Times Series and Forecasting, Second Edition
Carmona: Statistical Analysis of Financial Data in S-Plus
Chow and Teicher: Probability Theory: Independence, Interchangeability, Martingales, Third Edition
Christensen: Advanced Linear Modeling: Multivariate, Time Series, and Spatial Data; Nonparametric Regression and Response Surface Maximization, Second Edition
Christensen: Log-Linear Models and Logistic Regression, Second Edition
Christensen: Plane Answers to Complex Questions: The Theory of Linear Models, Third Edition
Creighton: A First Course in Probability Models and Statistical Inference
Davis: Statistical Methods for the Analysis of Repeated Measurements
Dean and Voss: Design and Analysis of Experiments
du Toit, Steyn, and Stumpf: Graphical Exploratory Data Analysis
Durrett: Essentials of Stochastic Processes
Edwards: Introduction to Graphical Modelling, Second Edition
Finkelstein and Levin: Statistics for Lawyers
Flury: A First Course in Multivariate Statistics
Jobson: Applied Multivariate Data Analysis, Volume I: Regression and Experimental Design
Jobson: Applied Multivariate Data Analysis, Volume II: Categorical and Multivariate Methods
Kalbfleisch: Probability and Statistical Inference, Volume I: Probability, Second Edition
Kalbfleisch: Probability and Statistical Inference, Volume II: Statistical Inference, Second Edition
Karr: Probability
Keyfitz: Applied Mathematical Demography, Second Edition
Kiefer: Introduction to Statistical Inference
Kokoska and Nevison: Statistical Tables and Formulae
Kulkarni: Modeling, Analysis, Design, and Control of Stochastic Systems
Lange: Applied Probability
Lange: Optimization
Lehmann: Elements of Large-Sample Theory
Lehmann: Testing Statistical Hypotheses, Second Edition

(continued after index)

Kenneth Lange

Optimization

Kenneth Lange
Department of Biomathematics and Human Genetics
UCLA School of Medicine
Box 951766
Los Angeles, CA 90095-1766
USA

Library of Congress Cataloging in Publication Data
Lange, Kenneth.
Optimization / Kenneth Lange.
p. cm. --(Springer Texts in Statistics)
Includes bibliographical references and index.
ISBN 0-387-20332-X (hard cover : alk. paper)

[on file]

ISBN 0-387-20332-X Printed on acid-free paper.

Printed in the United States of America. (MP)

9 8 7 6 5 4 3 2 1 SPIN 10951132

Springer-Verlag is a part of *Springer Science+Business Media*

springeronline.com

To my daughters, Jane and Maggie

Preface

This foreword, like many forewords, was written afterwards. That is just as well because the plot of the book changed during its creation. It is painful to recall how many times classroom realities forced me to shred sections and start anew. Perhaps such adjustments are inevitable. Certainly I gained a better perspective on the subject over time. I also set out to teach optimization theory and wound up teaching mathematical analysis. The students in my classes are no less bright and eager to learn about optimization than they were a generation ago, but they tend to be less prepared mathematically. So what you see before you is a compromise between a broad survey of optimization theory and a textbook of analysis. In retrospect, this compromise is not so bad. It compelled me to revisit the foundations of analysis, particularly differentiation, and to get right to the point in optimization theory.

The content of courses on optimization theory varies tremendously. Some courses are devoted to linear programming, some to nonlinear programming, some to algorithms, some to computational statistics, and some to mathematical topics such as convexity. In contrast to their gaps in mathematics, most students now come well trained in computing. For this reason, there is less need to emphasize the translation of algorithms into computer code. This does not diminish the importance of algorithms, but it does suggest putting more stress on their motivation and theoretical properties. Fortunately, the dichotomy between linear and nonlinear programming is fading. It makes better sense pedagogically to view linear programming as a special case of nonlinear programming. This is the attitude taken in the current book, which makes little mention of the simplex method and develops interior point methods instead. The real bridge between linear and nonlinear programming is convexity. I stress not only the theoretical side of convexity but also its applications in the design of algorithms for problems with either large numbers of parameters or nonlinear constraints.

This graduate-level textbook presupposes knowledge of calculus and linear algebra. I develop quite a bit of mathematical analysis from scratch and feature a variety of examples from linear algebra, differential equations, and convexity theory. Of course, the greater the prior exposure of students to this background material, the more quickly the beginning chapters can be covered. If the need arises, I recommend the texts [36, 65, 66, 92, 108, 109] for supplementary reading. There is ample material here for a fast-paced, semester-long course. Instructors should exercise their own discretion in skipping sections or chapters. For example, Chapter 8 on the EM algorithm primarily serves the needs of students in biostatistics and statistics. Overall, my intended audience includes graduate students in applied mathematics, biostatistics, computational biology, computer science, economics, physics, and statistics. To this list I would like to add upper-division majors in mathematics who want to see some rigorous mathematics with real applications. My own background in computational biology and statistics has obviously dictated many of the examples in the book.

Chapter 1 starts with a review of exact methods for solving optimization problems. These are methods that many students will have seen in calculus, but repeating classical techniques with fresh examples tends simultaneously to entertain, instruct, and persuade. Some of the exact solutions also appear later in the book as parts of more complicated algorithms.

Chapters 2 and 3 review undergraduate mathematical analysis. Although much of this material is standard, the examples may keep the interest of even the best students. Instructors should note that Carathéodory's definition rather than Fréchet's definition of differentiability is adopted. This choice eases the proof of many results. The gauge integral, another good addition to the calculus curriculum, is mentioned briefly.

Chapter 4 gets down to the serious business of optimization theory. McShane's clever proof of the necessity of the Karush-Kuhn-Tucker conditions avoids the complicated machinery of manifold theory and convex cones. It makes immediate use of the Mangasarian-Fromovitz constraint qualification. To derive sufficient conditions for optimality, I introduce second differentials by extending Carathéodory's definition of first differentials. To my knowledge, this approach to second differentials is new. Because it melds so effectively with second-order Taylor expansions, it renders critical proofs more transparent.

Chapter 5 treats convex sets, convex functions, and the relationship between convexity and the multiplier rule. The chapter concludes with the derivation of some of the classical inequalities of probability theory. Prior exposure to probability theory will obviously be an asset for readers here.

Chapters 6 and 7 introduce the MM and EM algorithms. These exploit convexity and the notion of majorization in transferring minimization of the objective function to a surrogate function. Minimizing the surrogate function drives the objective function downhill. The EM algorithm, which is a special case of the MM algorithm, arose in statistics. It is a slight misnomer to call these algorithms. They are really prescriptions for constructing algorithms. It takes experience and skill to wield these tools effectively, so careful attention to the examples is imperative.

Chapter 8 covers Newton's method and its statistical variants, scoring and the Gauss-Newton algorithm. To make this material less dependent on statistical knowledge, I have tried to motivate several algorithms from the perspective of positive definite approximation of the second differential of the objective function. Chapter 9 covers the conjugate gradient algorithm, quasi-Newton algorithms, and the method of trust regions. These classical subjects are in danger of being dropped from the curriculum of nonlinear programming. In my view, this would be a mistake.

Chapter 10 is devoted to convergence questions, both local and global. This material beautifully illustrates the virtues of soft analysis. Instructors wanting to emphasize practical matters may be tempted to sacrifice Chapter 10, but the constant interplay between theory and practice in designing new algorithms argues for its inclusion.

Chapter 11 on convex programming ends the book where more advanced treatises would start. I discuss adaptive barrier methods as a novel application of the MM algorithm, Dykstra's algorithm for finding feasible points in convex programming, and the rudiments of duality theory. These topics belong to the promised land. All you get here is a glimpse from the mountaintop looking out across the river.

Let me add a few words about notation. Lower-division undergraduate texts carefully distinguish between scalars and vectors by setting vectors in boldface type. This convention is considered cumbersome in higher mathematics and is dropped. However, mathematical analysis is plagued by a proliferation of superscripts and subscripts. I prefer to avoid superscripts because of the possible confusion with powers. This decision makes it difficult to distinguish an element of a vector sequence from a component of a vector. My compromise is to represent the mth entry of a vector sequence as $x_{(m)}$ and the nth component of that sequence element as x_{mn}. Similar conventions hold for matrices. Thus, M_{jkl} is the entry in row k and column l of the jth element $M_{(j)}$ of a sequence of matrices. Elements of scalar sequences are subscripted in the usual fashion without the enclosing parentheses.

I would like to thank my UCLA students for their help and patience in debugging this text. If it is readable, it is because their questions cut through the confusion. In retrospect, there were more contributing students than I can credit. Let me single out Jason Aten, Lara Bauman, Brian Dolan, Wei-Hsun Liao, Andrew Nevai-Tucker, Robert Rovetti, and Andy Yip. Paul

Maranian kindly prepared the index and proofread my last draft. Finally, I thank my ever helpful and considerate editor, John Kimmel.

I dedicate this book to my daughters, Jane and Maggie. It has been a privilege to be your father. Now that you are adults, I hope you can find the same pleasure in pursuing ideas that I have found in my professional life.

Contents

1
Elementary Optimization

1.1 Introduction

Optimization theory is one of the oldest branches of mathematics, serving as a catalyst for the development of geometry and differential calculus [117]. Today it finds applications in a myriad of scientific and engineering disciplines. The current chapter briefly surveys material that most students encounter in a good calculus course. This review is intended to showcase the variety of methods used to find the exact solutions of elementary problems. We will return to some of these methods later from a more rigorous perspective. One of the recurring themes in optimization theory is its close connection to inequalities. This chapter introduces a few classical inequalities; more will appear in succeeding chapters.

1.2 Univariate Optimization

The first optimization problems students encounter are univariate. Solution techniques for these simple problems are hardly limited to differential calculus. Our first two examples illustrate how plane geometry and algebra can play a role.

Example 1.2.1 *Heron's Problem*

The ancient mathematician Heron of Alexandria posed one of the earliest optimization problems. Consider the two points A and B and the line

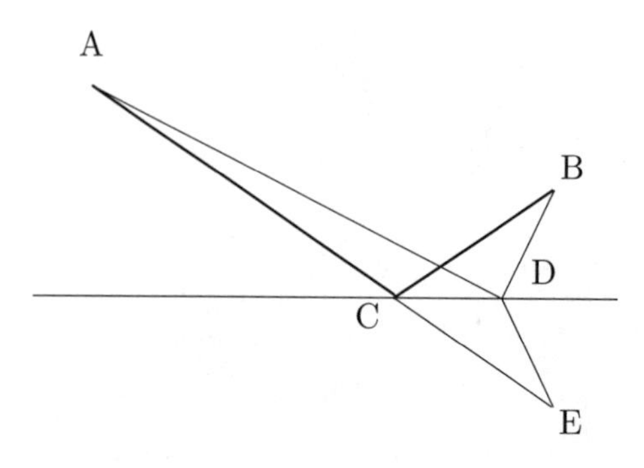

FIGURE 1.1. Diagram for Heron's Problem

containing the points C and D drawn in Figure 1.1. Heron's problem is to find the position of C on the line that minimizes the sum of the distances $|AC|$ and $|BC|$. The correct choice of C is determined by reflecting B across the given line to give E. From E we draw the line to A and note its intersection C with the original line. To demonstrate that C minimizes the total distance $|AC|+|BC|$, consider any other point D on the original horizontal line. By symmetry, $|AC|+|BC| = |AC|+|CE|$. Similarly, by symmetry, $|AD|+|BD| = |AD|+|DE|$. Because the sum of the lengths of two sides of a triangle exceeds the length of the third side, it follows immediately that $|AC|+|CE| \leq |AD|+|DE|$. Thus, C solves Heron's problem.

This example also has an optical interpretation. If we imagine that the horizontal line containing C lies on a mirror, then light travels along the quickest path between A and B via the mirror. This extremal principle can be explained by considering the wave nature of light, but we omit the long digression. It is interesting that the geometric argument automatically implies that the angle of incidence of the light ray equals the angle of reflection. ■

Example 1.2.2 *Simple Arithmetic-Geometric Mean Inequality*

If x and y are two nonnegative numbers, then $\sqrt{xy} \leq (x+y)/2$. This can be proved by noting that

$$\begin{aligned} 0 &\leq (\sqrt{x}-\sqrt{y})^2 \\ &= x - 2\sqrt{xy} + y. \end{aligned}$$

Evidently, equality holds if and only if $x = y$. As an application consider maximization of the function $f(x) = x(1-x)$. The inequality just derived shows that

$$f(x) \leq \left(\frac{x+1-x}{2}\right)^2 = \frac{1}{4},$$

with equality when $x = 1/2$. Thus, the maximum of $f(x)$ occurs at the point $x = 1/2$. One can interpret $f(x)$ as the area of a rectangle of fixed perimeter 2 with sides of length x and $1-x$. The rectangle with the largest area is a square. The function $2f(x)$ is interpreted in population genetics as the fraction of a population that is heterozygous at a genetic locus with two alleles having frequencies x and $1 - x$. Heterozygosity is maximized when the two alleles are equally frequent. ■

With the advent of differential calculus, it became possible to solve optimization problems more systematically. Before discussing concrete examples, it is helpful to review some of the standard theory. We restrict attention to real-valued functions defined on intervals. The intervals in question can be finite or infinite in extent and open or closed at either end. According to a celebrated theorem of Weierstrass, a continuous function $f(x)$ defined on a closed finite interval $[a, b]$ attains its minimum and maximum values on the interval. These extremal values are necessarily finite. The extremal points can occur at the endpoints a or b or at an interior point c. In the later case, when $f(x)$ is differentiable, an even older principle of Fermat requires that $f'(c) = 0$. The stationarity condition $f'(c) = 0$ is no guarantee that c is optimal. It is possible for c to be a local rather than a global minimum or maximum or even to be a saddle point. However, it usually is a simple matter to check the endpoints a and b and any stationary points c. Collectively, these points are known as critical points.

If the domain of $f(x)$ is not a closed finite interval $[a, b]$, then the minimum or maximum of $f(x)$ may not exist. One can usually rule out such behavior by examining the limit of $f(x)$ as x approaches an open boundary. For example on the interval $[a, \infty)$, if $\lim_{x \to \infty} f(x) = \infty$, then we can be sure that $f(x)$ possesses a minimum on the interval, and we can find it by comparing the values of $f(x)$ at a and any stationary points c. On a half open interval such as $(a, b]$, we can likewise find a minimum whenever $\lim_{x \to a} f(x) = \infty$. Similar considerations apply to finding a maximum.

The nature of a stationary point c can be determined by testing the second derivative $f''(c)$. If $f''(c) > 0$, then c at least qualifies as a local minimum. Similarly, if $f''(c) < 0$, then c at least qualifies as a local maximum. The indeterminate case $f''(c) = 0$ is consistent with c being a local minimum, maximum, or saddle point. For example, $f(x) = x^4$ attains its minimum at 0 while $f(x) = x^3$ has a saddle point there. In both cases, $f''(0) = 0$. Higher-order derivatives or other qualitative features of $f(x)$ must be invoked to discriminate among these possibilities. If $f''(x) \geq 0$ for all x, then $f(x)$ is said to be convex. Any stationary point of a convex function is a minimum. If $f''(x) > 0$ for all x, then $f(x)$ is strictly convex, and there is at most one stationary point. Whenever it exists, the stationary point furnishes the global minimum. A concave function satisfies $f''(x) \leq 0$ for all x. Concavity bears the same relation to maxima as convexity does to minima.

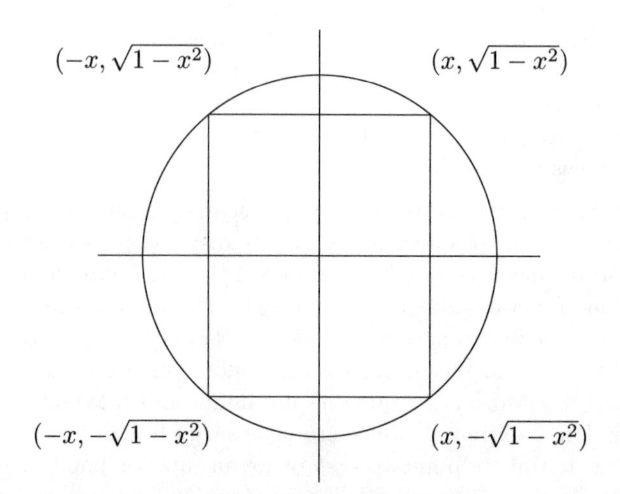

FIGURE 1.2. A Rectangle Inscribed in a Circle

Example 1.2.3 (Kepler) *Largest Rectangle Inscribed in a Circle*

Figure 1.2 depicts a rectangle inscribed in a circle of radius 1 centered at the origin. If we suppose the vertical sides of the rectangle cross the horizontal axis at the point $(-x, 0)$ and $(x, 0)$, then Pythagoras's theorem gives the coordinates of the corners as noted in the figure. Here x is restricted to the interval $[0, 1]$. From these coordinates, it follows that the rectangle has area

$$f(x) = 4x\sqrt{1-x^2}.$$

Because $f(0) = f(1) = 0$, the maximum of $f(x)$ occurs somewhere in the open interval $(0, 1)$. Straightforward differentiation shows that

$$f'(x) = 4\sqrt{1-x^2} - \frac{4x^2}{\sqrt{1-x^2}}.$$

Setting $f'(x)$ equal to 0 and solving for x gives the critical point $x = 1/\sqrt{2}$ and the critical value $f(1/\sqrt{2}) = 2$. Since there is only one critical point on $(0, 1)$, it must be the maximum point. The largest inscribed rectangle is a square as expected. ■

Example 1.2.4 *Snell's Law*

Snell's law refers to an optical experiment involving two different media, say air and water. The less dense the medium, the faster light travels. Since light takes the path of least time, it bends at an interface such as that indicated by the horizontal axis in Figure 1.3. Here we ask for the point $(x, 0)$ on the interface intersecting the light path. If we assume the speed of

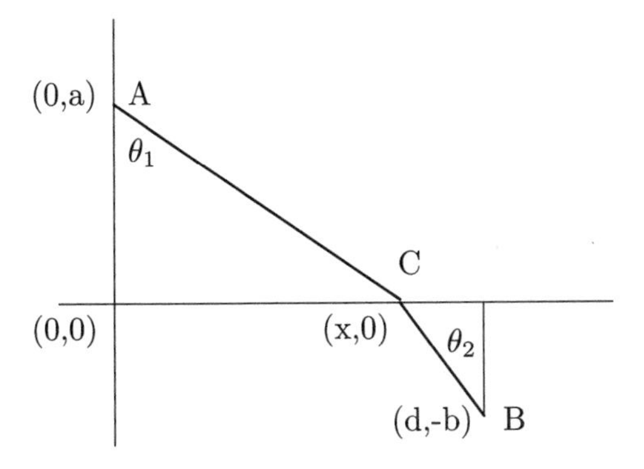

FIGURE 1.3. Diagram for Snell's Law

light above the interface is s_1 and below the interface is s_2, then the total travel time is given by

$$f(x) = \frac{\sqrt{a^2 + x^2}}{s_1} + \frac{\sqrt{b^2 + (d - x)^2}}{s_2}.$$

The derivative of $f(x)$ is

$$f'(x) = \frac{x}{s_1\sqrt{a^2 + x^2}} - \frac{d - x}{s_2\sqrt{b^2 + (d - x)^2}}.$$

The minimum exists because $\lim_{|x|\to\infty} f(x) = \infty$. Although finding a stationary point is difficult, it is clear from the monotonicity of the functions $x/\left(s_1\sqrt{a^2 + x^2}\right)$ and $(d-x)/\left(s_2\sqrt{b^2 + (d - x)^2}\right)$ that it is unique. In trigonometric terms, Snell's law can be expressed as

$$\frac{\sin\theta_1}{s_1} = \frac{\sin\theta_2}{s_2}$$

using the angles at the minimum point as noted in Figure 1.3. ■

Example 1.2.5 *The Functions* $f_n(x) = x^n e^x$

The functions $f_n(x) = x^n e^x$ for $n \geq 1$ exhibit interesting behavior. Figure 1.4 plots $f_n(x)$ for n between 1 and 3. It is clear that $\lim_{x\to-\infty} f_n(x) = 0$ and $\lim_{x\to\infty} f_n(x) = \infty$. These limits do not rule out the possibility of local maxima and minima. To find these we need

$$\begin{aligned} f_n'(x) &= (x^n + nx^{n-1})e^x \\ f_n''(x) &= [x^n + 2nx^{n-1} + n(n-1)x^{n-2}]e^x. \end{aligned}$$

Setting $f_n'(x) = 0$ produces the critical point $x = -n$, and when $n > 1$, the critical point $x = 0$. A brief calculation shows that $f_n''(-n) = (-n)^{n-1}e^{-n}$.

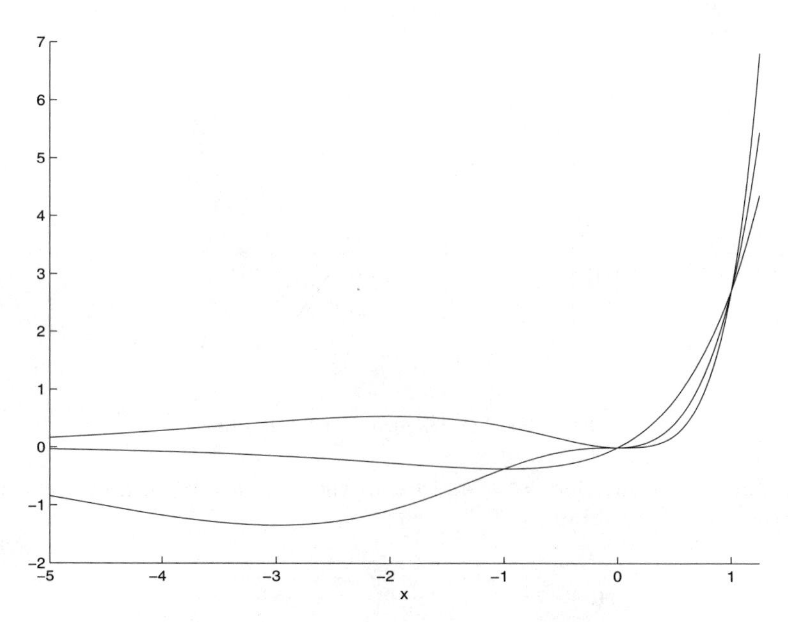

FIGURE 1.4. Plots of xe^x, x^2e^x, and x^3e^x

Thus, $-n$ is a local minimum for n odd and a local maximum for n even. At 0 we have $f_2''(0) = 2$ and $f_n''(0) = 0$ for $n > 2$. Thus, the second derivative test fails for $n > 2$. However, it is clear from the variation of the sign of $f_n(x)$ to the right and left of 0 that 0 is a minimum of $f_n(x)$ for n even and a saddle point of $f_n(x)$ for $n > 1$ and odd. One strength of modern graphing programs such as MATLAB is that they quickly suggest such conjectures. ■

Example 1.2.6 *Fenchel Conjugate of $f_p(x) = |x|^p/p$ for $p > 1$*

The Fenchel conjugate $f^\star(y)$ of a convex function $f(x)$ is defined by

$$f^\star(y) = \sup_x \{yx - f(x)\}.$$

Remarkably, $f^\star(y)$ is also convex. As a particular case of this result, we consider the Fenchel conjugate of $f_p(x)$. It turns out that $f_p^\star(y) = f_q(y)$, where

$$\frac{1}{p} + \frac{1}{q} = 1.$$

Here neither p nor q need be integers. According to the second derivative test, the function $f_p(x) = |x|^p/p$ is convex on the real line whenever $p > 1$. The possible failure of $f_p''(x)$ to exist at $x = 0$ does not invalidate this conclusion. To calculate $f_p^\star(y)$, we observe that $f_p'(x) = |x|^{p-1}\,\mathrm{sgn}(x)$. This clearly implies that $x = |y|^{1/(p-1)} \operatorname{sgn}(y)$ maximizes the concave function $g(x) = yx - f_p(x)$. At the maximum point

$$\begin{aligned} f_p^\star(y) &= yx - \frac{|x|^p}{p} \\ &= |y|^{1+1/(p-1)} - \frac{|y|^{p/(p-1)}}{p} \\ &= \frac{|y|^q}{q}, \end{aligned}$$

proving our claim. The conclusion $f_p^{\star\star}(x) = f_p(x)$ is a special case of the general result proved later in Example 11.4.4. Historically, the Fenchel conjugate was introduced by Legendre for smooth functions and later generalized by Fenchel to arbitrary functions. ■

1.3 Multivariate Optimization

Although multivariate optimization is more subtle, it typically parallels univariate optimization [61, 104, 113]. The most fundamental differences arise because of constraints. In unconstrained optimization, the right definitions and notation ease the generalization. Before discussing these issues of calculus, we look at two classical inequalities that can be established by purely algebraic techniques.

Example 1.3.1 *Cauchy-Schwarz Inequality*

Suppose x and y are any two points in R^n. The Cauchy-Schwarz inequality says

$$\left|\sum_{i=1}^n x_i y_i\right| \leq \left(\sum_{i=1}^n x_i^2\right)^{1/2} \left(\sum_{i=1}^n y_i^2\right)^{1/2}.$$

If we define the inner product

$$x^* y = \sum_{i=1}^n x_i y_i$$

using the transpose operator * and the Euclidean norm

$$\|x\| = \left(\sum_{i=1}^n x_i^2\right)^{1/2},$$

then the inequality can be restated as $|x^* y| \leq \|x\| \cdot \|y\|$. Equality occurs in the Cauchy-Schwarz inequality if and only if y is a multiple of x or vice versa.

In proving the inequality, we can immediately eliminate the case $x = \mathbf{0}$ where all components of x are 0. Given that $x \neq \mathbf{0}$, we introduce a scalar λ and consider the quadratic

$$\begin{aligned} 0 &\leq \|\lambda x + y\|^2 \\ &= \|x\|^2\lambda^2 + 2x^*y\lambda + \|y\|^2 \\ &= \frac{1}{a}(a\lambda + b)^2 + c - \frac{b^2}{a} \end{aligned}$$

with $a = \|x\|^2$, $b = x^*y$, and $c = \|y\|^2$. In order for this quadratic to be nonnegative for all λ, it is necessary and sufficient that $c - b^2/a \geq 0$, which is just an abbreviation for the Cauchy-Schwarz inequality. For the quadratic to attain the value 0, the condition $c - b^2/a = 0$ must hold. When the quadratic vanishes, $y = -\lambda x$. ■

Example 1.3.2 *General Arithmetic-Geometric Mean Inequality*

One generalization of the simple arithmetic-geometric mean inequality of Example 1.2.2 takes the form

$$\sqrt[n]{x_1 \cdots x_n} \leq \frac{x_1 + \cdots + x_n}{n}, \tag{1.1}$$

where $x_1, \ldots, x_n$ are any n nonnegative numbers. For a purely algebraic proof of this fact, we first note that it is obvious if any $x_i = 0$. If all $x_i > 0$, then divide both sides of the inequality by $\sqrt[n]{x_1 \cdots x_n}$. This replaces x_i by $y_i = x_i / \sqrt[n]{x_1 \cdots x_n}$ and leads to the equality $\sqrt[n]{y_1 \cdots y_n} = 1$. It now suffices to prove that $y_1 + \cdots + y_n \geq n$, which is trivially valid when $n = 1$. For $n > 1$ we argue by induction. Clearly the assumption $\sqrt[n]{y_1 \cdots y_n} = 1$ implies that there are two numbers, say y_1 and y_2, with $y_1 \geq 1$ and $y_2 \leq 1$. If this is true, then $(y_1 - 1)(y_2 - 1) \leq 0$, or equivalently $y_1 y_2 + 1 \leq y_1 + y_2$. Invoking the induction hypothesis, we now reason that

$$\begin{aligned} y_1 + \cdots + y_n &\geq 1 + y_1 y_2 + y_3 + \cdots + y_n \\ &\geq 1 + (n-1). \end{aligned}$$

■

As a prelude to discussing further examples, it is helpful to briefly summarize the theory to be developed later and often taken for granted in multidimensional calculus courses. The standard vocabulary and symbolism adopted here stress the minor adjustments necessary in going from one dimension to multiple dimensions.

For a real-valued function $f(x)$ defined on R^n, the differential $df(x)$ is the generalization of the derivative $f'(x)$. For our purposes, $df(x)$ is the row vector of partial derivatives; its transpose is the gradient vector $\nabla f(x)$. The symmetric matrix of second partial derivatives constitutes the second differential $d^2 f(x)$ or Hessian matrix. A stationary point x satisfies

$\nabla f(x) = \mathbf{0}$. Fermat's principle says that all local maxima and minima on the interior of the domain of $f(x)$ are stationary points.

If $d^2 f(x)$ is positive definite at a stationary point x, then x furnishes a local minimum. If $d^2 f(x)$ is negative definite, then x furnishes a local maximum. The function $f(x)$ is said to be convex if $d^2 f(x)$ is positive semidefinite for all x; it is strictly convex if $d^2 f(x)$ is positive definite for all x. Every stationary point of a convex function represents a global minimum. At most one stationary point exists per strictly convex function. Similar considerations apply to concave functions and global maxima, provided we substitute "negative" for "positive" throughout these definitions.

Example 1.3.3 *Least Squares Estimation*

Statisticians often estimate parameters by the method of least squares. To review the situation, consider m independent experiments with outcomes $y_1, \ldots, y_m$. We wish to predict y_i from n covariates $x_{i1}, \ldots, x_{in}$ known in advance. For instance, y_i might be the height of the ith child in a classroom of m children. Relevant covariates might be the heights x_{i1} and x_{i2} of i's mother and father and the sex of i coded as $x_{i3} = 1$ for a girl and $x_{i4} = 1$ for a boy. Here we take $n = 4$ and force $x_{i3}x_{i4} = 0$ so that only one sex is possible. If we use a linear predictor $\sum_{j=1}^n x_{ij}\theta_j$ of y_i, it is natural to estimate the regression coefficients θ_j by minimizing the sum of squares

$$f(\theta) = \sum_{i=1}^m \Big(y_i - \sum_{j=1}^n x_{ij}\theta_j\Big)^2.$$

Differentiating $f(\theta)$ with respect to θ_j and setting the result equal to 0 produce

$$\sum_{i=1}^m x_{ij}y_i = \sum_{i=1}^m \sum_{k=1}^n x_{ij}x_{ik}\theta_k.$$

If we let y denote the column vector with entries y_i and X denote the matrix with entry x_{ij} in row i and column j, then these n normal equations can be written in vector form as

$$X^* y = X^* X\theta$$

and solved as

$$\hat{\theta} = (X^* X)^{-1} X^* y.$$

In order for the indicated matrix inverse $(X^*X)^{-1}$ to exist, $m \geq n$ should hold and the matrix X must be of full rank. See Problem 14.

To check that our proposed solution $\hat{\theta}$ represents the global minimum, we calculate the Hessian matrix $d^2 f(\theta)$. Its entries

$$\frac{\partial^2}{\partial\theta_j \partial\theta_k} f(\theta) = 2\sum_{i=1}^m x_{ij}x_{ik}$$

permit us to identify $d^2f(\theta)$ with the matrix $2X^*X$. Owing to the full rank assumption, the symmetric matrix X^*X is positive definite. Hence, $f(\theta)$ is strictly convex, and $\hat{\theta}$ is the global minimum. ■

1.4 Constrained Optimization

The subject of Lagrange multipliers has a strong geometric flavor. It deals with tangent vectors and directions of steepest ascent and descent. The classical theory, which is all we consider here, is limited to equality constraints. Inequality constraints were not introduced until later in the game.

The gradient direction $\nabla f(x) = df(x)^*$ is the direction of steepest ascent of $f(x)$ near the point x. We can motivate this fact by considering the linear approximation

$$f(x + tu) = f(x) + tdf(x)u + o(t)$$

for a unit vector u and a scalar t. The error term $o(t)$ becomes negligible compared to t as t decreases to 0. The inner product $df(x)u$ in this approximation is greatest for the unit vector $u = \nabla f(x)/\|\nabla f(x)\|$. Thus, $\nabla f(x)$ points locally in the direction of steepest ascent of $f(x)$. Similarly, $-\nabla f(x)$ points locally in the direction of steepest descent.

Now consider minimizing or maximizing $f(x)$ subject to the equality constraints $g_i(x) = 0$ for $i = 1, \ldots, m$. A tangent direction w at the point x on the constraint surface satisfies $dg_i(x)w = 0$ for all i. Of course, if the constraint surface is curved, we must interpret the tangent directions as specifying directions of infinitesimal movement. From the perpendicularity relation $dg_i(x)w = 0$, it follows that the set of tangent directions is the orthogonal complement $S^\perp(x)$ of the vector subspace $S(x)$ spanned by the $\nabla g_i(x)$. To avoid degeneracies, the vectors $\nabla g_i(x)$ must be linearly independent. Figure 1.5 depicts level curves $g(x) = c$ and gradients $\nabla g(x)$ for the function $\sin(x)\cos(y)$ over the square $[0, \pi] \times [-\frac{\pi}{2}, \frac{\pi}{2}]$. Tangent vectors are parallel to the level curves (contours) and perpendicular to the gradients (arrows).

At an optimal (or extremal) point y, we have $df(y)w = 0$ for every tangent direction $w \in S^\perp(y)$; otherwise, we could move infinitesimally away from y in the tangent directions w and $-w$ and both increase and decrease $f(x)$. In other words, $\nabla f(y)$ is a member of the double orthogonal complement $S^{\perp\perp}(y) = S(y)$. This enables us to write

$$\nabla f(y) = -\sum_{i=1}^{m} \lambda_i \nabla g_i(y)$$

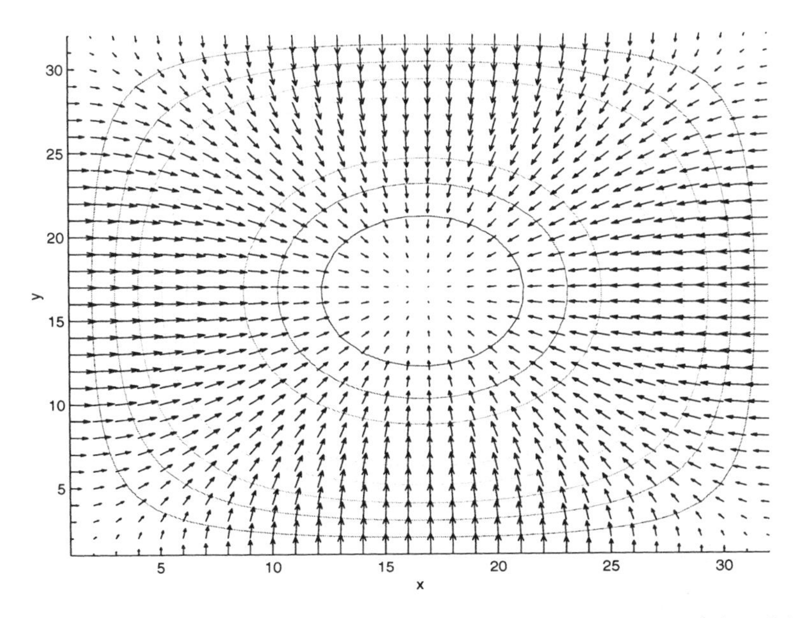

FIGURE 1.5. Level Curves and Steepest Ascent Directions for $\sin(x)\cos(y)$

for properly chosen constants $\lambda_1, \ldots, \lambda_m$. Alternatively, the Lagrangian function

$$\mathcal{L}(x, \omega) = f(x) + \sum_{i=1}^{m} \omega_i g_i(x)$$

has a stationary point at (y, λ). In this regard, note that

$$\frac{\partial}{\partial \omega_i} \mathcal{L}(y, \lambda) = 0$$

owing to the constraint $g_i(y) = 0$. The essence of the Lagrange multiplier rule consists in finding a stationary point of the Lagrangian. Although our intuitive arguments need logical tightening in many places, they offer the basic geometric insights.

Example 1.4.1 *Projection onto a Hyperplane*

A hyperplane in R^n is the set of points $H = \{x \in \mathsf{R}^n : z^*x = c\}$ for some vector $z \in \mathsf{R}^n$ and scalar c. There is no loss in generality in assuming that z is a unit vector. If we seek the closest point on H to a point y, then we must minimize $\|y - x\|^2$ subject to $x \in H$. We accordingly form the Lagrangian

$$\mathcal{L}(x, \lambda) = \|y - x\|^2 + \lambda(z^*x - c).$$

Setting the partial derivative with respect to x_i equal to 0 gives

$$-2(y_i - x_i) + \lambda z_i = 0.$$

This equality entails $x = y - \frac{1}{2}\lambda z$ in vector notation. It follows that

$$c = z^* x = z^* y - \frac{1}{2}\lambda \|z\|^2.$$

In view of the assumption $\|z\| = 1$, we find that

$$\lambda = -2(c - z^* y)$$

and consequently that

$$x = y + (c - z^* y)z.$$

If $y \in H$ to begin with, then $x = y$. ■

Example 1.4.2 *Estimation of Multinomial Proportions*

As another statistical example, consider a multinomial experiment with m trials and observed successes $m_1, \ldots, m_n$ over n categories. The maximum likelihood estimate of the probability p_i of category i is $\hat{p}_i = m_i/m$, where $m = m_1 + \cdots + m_n$. To demonstrate this fact, let

$$L(p) = \binom{m}{m_1, \ldots, m_n} \prod_{i=1}^n p_i^{m_i}$$

denote the likelihood. If $m_i = 0$ for some i, then we interpret $p_i^{m_i}$ as 1 even when $p_i = 0$. This convention makes it clear that we can increase $L(p)$ by replacing p_i by 0 and p_j by $p_j/(1 - p_i)$ for $j \neq i$. Thus, for purposes of maximum likelihood estimation, we can assume that all $m_i > 0$. Given this assumption, $L(p)$ tends to 0 when any p_i tends to 0. It follows that we can further restrict our attention to the interior region where all $p_i > 0$ and maximize the loglikelihood $\ln L(p)$ subject to the equality constraint $\sum_{i=1}^n p_i = 1$. To find the maximum of $\ln L(p)$, we look for a stationary point of the Lagrangian

$$\mathcal{L}(p, \lambda) = \ln \binom{m}{m_1, \ldots, m_n} + \sum_{i=1}^n m_i \ln p_i + \lambda \Big(\sum_{i=1}^n p_i - 1 \Big).$$

Setting the partial derivative of $\mathcal{L}(p, \lambda)$ with respect to p_i equal to 0 gives the equation

$$-\frac{m_i}{p_i} = \lambda.$$

These n equations are satisfied subject to the constraint by taking $\lambda = -m$ and $\hat{p}_i = m_i/m$. Thus, the necessary condition for a maximum holds at $\hat{p}$. One can show that $\hat{p}$ furnishes the global maximum by exploiting the strict concavity of $L(p)$. Although we will omit the details of this argument, it is fair to point out that strict concavity follows from

$$\frac{\partial^2}{\partial p_i \partial p_j} \ln L(p) = \begin{cases} -\frac{m_i}{p_i^2} & i = j \\ 0 & i \neq j\,. \end{cases}$$

In statistical applications, the negative second differential $-d^2 \ln L(p)$ is called the observed information matrix. ■

Example 1.4.3 *Eigenvalues of a Symmetric Matrix*

Let $M = (m_{ij})$ be an $n \times n$ symmetric matrix. Recall that M has n real eigenvalues and n corresponding orthogonal eigenvectors. To find the minimum or maximum eigenvalue of M, consider optimizing the quadratic form x^*Mx subject to the constraint $\|x\|^2 = 1$. To handle this nonlinear constraint, we introduce the Lagrangian

$$\mathcal{L}(x, \lambda) = x^*Mx + \lambda(\|x\|^2 - 1).$$

Setting the partial derivative of $\mathcal{L}(x, \lambda)$ with respect to x_i equal to 0 yields

$$2\sum_{j=1}^{n} m_{ij}x_j + 2\lambda x_i = 0.$$

In matrix notation, this reduces to $Mx = -\lambda x$. It follows that

$$x^*Mx = -\lambda x^*x = -\lambda.$$

Thus, the stationary points of the Lagrangian are eigenvectors of M. The negative Lagrange multipliers are the corresponding stationary values or eigenvalues. The maximum and minimum eigenvalues occur among these stationary values. ■

Example 1.4.4 *A Population Genetics Problem*

The multiplier rule is sometimes hard to apply, and ad hoc methods can lead to better results. In the setting of the multinomial distribution, consider the problem of maximizing the sum

$$f(p) = \sum_{i<j} (2p_i p_j)^2 \frac{1}{2}$$

subject to the constraints $\sum_{i=1}^{n} p_i = 1$ and all $p_i \geq 0$. This problem has a genetics interpretation involving a locus with n codominant alleles labeled

$1, \ldots, n$. (See Section 6.4 for some genetics terminology.) At the locus, most genotype combinations of a mother, father, and child make it possible to infer which allele the mother contributes to the child and which allele the father contributes to the child. It turns out that the only ambiguous case occurs when all three family members share the same heterozygous genotype i/j, where $i \neq j$. The probability of this configuration is $(2p_ip_j)^2\frac{1}{2}$ if p_i and p_j are the population frequencies (proportions) of alleles i and j. Here $2p_ip_j$ is the frequency of an i/j mother or an i/j father and $\frac{1}{2}$ is the probability that one of them transmits an i allele and the other transmits a j allele. Thus, $f(p)$ represents the probability that the trio's genotypes do not permit inference of the child's maternal and paternal alleles.

The case $n = 2$ is particularly simple because the function $f(p)$ then reduces to $2(p_1p_2)^2$. In view of Example 1.2.2, the maximum of $\frac{1}{8}$ is attained when $p_1 = p_2 = \frac{1}{2}$. This suggests that the maximum for general n occurs when all $p_i = \frac{1}{n}$. Because there are $\binom{n}{2}$ heterozygous genotypes,

$$\begin{aligned} f\Big(\frac{1}{n}\mathbf{1}\Big) &= \binom{n}{2}\Big(\frac{2}{n^2}\Big)^2\frac{1}{2} \\ &= \frac{n-1}{n^3}, \end{aligned}$$

which is strictly less than $\frac{1}{8}$ for $n \geq 3$. Our first guess is wrong, and we now conjecture that the maximum occurs on a boundary where all but two of the $p_i = 0$. If we permute the components of a maximum point, then symmetry dictates that the result will also be a maximum point. We therefore order the parameters so that $0 < p_1 \leq p_2 \leq \cdots \leq p_n$, avoiding for the moment the lower-dimensional case where $p_1 = 0$.

We now argue that we can increase $f(p)$ by increasing p_2 by $q \in [0, p_1]$ at the expense of decreasing p_1 by q. Consider the function

$$g(q) = 2(p_1 - q)^2 \sum_{i=3}^n p_i^2 + 2(p_2 + q)^2 \sum_{i=3}^n p_i^2 + 2(p_1 - q)^2(p_2 + q)^2$$

which equals the original objective function except for an additive constant independent of q. For $n \geq 3$, straightforward differentiation gives

$$\begin{aligned} g'(q) &= -4(p_1 - q)\sum_{i=3}^n p_i^2 + 4(p_2 + q)\sum_{i=3}^n p_i^2 \\ &\quad - 4(p_1 - q)(p_2 + q)^2 + 4(p_1 - q)^2(p_2 + q) \\ &= 4(p_2 - p_1 + 2q)\Big[\sum_{i=3}^n p_i^2 - (p_1 - q)(p_2 + q)\Big] \\ &\geq 4(p_2 - p_1 + 2q)\Big[\sum_{i=3}^n p_i^2 - (p_2 - q)(p_2 + q)\Big] \end{aligned}$$

$$
\begin{aligned}
&= 4(p_2 - p_1 + 2q)\Big[\sum_{i=4}^{n} p_i^2 + p_3^2 - p_2^2 + q^2\Big] \\
&\geq 0
\end{aligned}
$$

for $q \in [0, p_1]$. Thus, we should reduce p_1 to 0 and increase p_2 to $p_2 + p_1$. Furthermore, we should keep discarding the lowest positive p_j until all but two of the p_i equal 0. Finally, we set the remaining two p_i equal to $\frac{1}{2}$. This verifies our second conjecture. ■

1.5 Problems

1. Given a point C in the interior of an acute angle, find the points A and B on the sides of the angle such that the perimeter of the triangle ABC is as short as possible.

2. Find the minima of the functions

$$
\begin{aligned}
f(x) &= x \ln x \\
g(x) &= x - \ln x \\
h(x) &= x + \frac{1}{x}
\end{aligned}
$$

on $(0, \infty)$. Demonstrate rigorously that your solutions are indeed the minima.

3. For $t > 0$ prove that $e^x > x^t$ for all $x > 0$ if and only if $t < e$ [29].

4. Demonstrate that Euler's function $f(x) = x^2 - 1/\ln x$ possesses no local or global minima on either domain $(0, 1)$ or $(1, \infty)$.

5. Prove the harmonic-geometric mean inequality

$$
\frac{1}{\frac{1}{n}\left(\frac{1}{x_1} + \cdots + \frac{1}{x_n}\right)} \leq \sqrt[n]{x_1 \cdots x_n}
$$

for n positive numbers $x_1, \ldots, x_n$.

6. Heron's classical formula for the area of a triangle with sides of length a, b, and c is $\sqrt{s(s-a)(s-b)(s-c)}$, where $s = (a+b+c)/2$ is the semiperimeter. Show that the triangle of fixed perimeter with greatest area is equilateral.

7. Let $H_n = 1 + \frac{1}{2} + \cdots + \frac{1}{n}$. Verify the inequality $n\sqrt[n]{n+1} \leq n + H_n$ for any positive integer n (Putnam Competition, 1975).

8. Find the point in a triangle that minimizes the sum of the squared distances from the vertices. Show that this point is the intersection of the medians of the triangle.

9. Given an angle in the plane and a point in its interior, find the line that passes through the point and cuts off from the angle a triangle of minimal area. This triangle is determined by the vertex of the angle and the two points where the constructed line intersects the sides of the angle.

10. Consider an n-gon circumscribing the unit circle in R^2. Demonstrate that the n-gon has minimum area if and only if all of its n sides have equal length. (Hint: Let θ_m be the circular angle between the two points of tangency of sides m and $m+1$ [29]. Show that the area of the quadrilateral defined by the center of the circle, the two points of tangency, and the intersection of the two sides is given by $\tan \frac{\theta_m}{2}$.)

11. In forensic applications of genetics, the sum

$$s = 1 - 2\Big(\sum_{i=1}^{n} p_i^2\Big)^2 + \sum_{i=1}^{n} p_i^4$$

occurs [83]. Here the p_i are nonnegative and sum to 1. Prove rigorously that s attains its maximum $s_{\max} = 1 - \frac{2}{n^2} + \frac{1}{n^3}$ when all $p_i = \frac{1}{n}$. (Hint: To prove the claim about $s_{\max}$, note that without loss of generality one can assume $p_1 \le p_2 \le \cdots \le p_n$. If $p_i < p_{i+1}$, then s can be increased by replacing p_i and p_{i+1} by $p_i + x$ and $p_{i+1} - x$ for x positive and sufficiently small.)

12. Suppose that a and b are real numbers satisfying $0 < a < b$. Prove that the origin locally minimizes $f(x, y) = (y - ax^2)(y - bx^2)$ along every line $x = ht$ and $y = kt$ through the origin. Also show that $f(x, cx^2) < 0$ for $a < c < b$ and $x \neq 0$ and that $f(x, cx^2) > 0$ for $c < a$ or $c > b$ and $x \neq 0$. The origin therefore affords a local minimum along each line through the origin but not a local minimum in the wider sense.

13. Demonstrate that the function $x_1^2 + x_2^2(1 - x_1)^3$ has a unique stationary point in R^2, which is a local minimum but not a global minimum. Can this occur for a continuously differentiable function with domain R?

14. Suppose that the $m \times n$ matrix X has full rank and that $m \ge n$. Show that the $n \times n$ matrix X^*X is invertible and positive definite.

15. Consider two sets of positive numbers $x_1, \ldots, x_n$ and $\alpha_1, \ldots, \alpha_n$ such that $\sum_{i=1}^n \alpha_i = 1$. Prove the generalized arithmetic-geometric mean

inequality

$$\prod_{i=1}^{n} x_i^{\alpha_i} \leq \sum_{i=1}^{n} \alpha_i x_i$$

by minimizing $\sum_{i=1}^{n} \alpha_i x_i$ subject to the constraint $\prod_{i=1}^{n} x_i^{\alpha_i} = c$.

16. Find the rectangular box in R^3 of greatest volume having a fixed surface area.

17. Let $S(\mathbf{0}, r) = \{x \in \mathsf{R}^n : \|x\| = r\}$ be the sphere of radius r centered at the origin. For $y \in \mathsf{R}^n$, find the point of $S(r)$ closest to y.

18. Find the parallelepiped of maximum volume that can be inscribed in the ellipsoid

$$\frac{x_1^2}{a^2} + \frac{x_2^2}{b^2} + \frac{x_3^2}{c^2} = 1.$$

Assume that the parallelepiped is centered at the origin and has edges parallel to the coordinate axes.

19. Prove the arithmetic-quadratic mean inequality

$$\frac{1}{n}\sum_{i=1}^{n} x_i \leq \left(\frac{1}{n}\sum_{i=1}^{n} x_i^2\right)^{1/2}$$

for any nonnegative numbers $x_1, \ldots, x_n$.

20. A twice continuously differentiable function $f(x)$ on R^2 satisfies

$$\frac{\partial^2}{\partial x_1^2} f(x) + \frac{\partial^2}{\partial x_2^2} f(x) > 0$$

for all x. Prove that $f(x)$ has no local maxima [29]. An example of such a function is $f(x) = \|x\|^2 = x_1^2 + x_2^2$.

2

The Seven C's of Analysis

2.1 Introduction

The current chapter explains key concepts of mathematical analysis summarized by the six adjectives convergent, complete, closed, compact, continuous, and connected. A later chapter will add to these six c's the seventh c, convex. At first blush these concepts seem remote from practical problems of optimization. However, painful experience and exotic counterexamples have taught mathematicians to pay attention to details. Fortunately, we can benefit from the struggles of earlier generations and bypass many of the intellectual traps.

2.2 Vector and Matrix Norms

In multidimensional calculus, vector and matrix norms quantify notions of topology and convergence [18, 48, 55, 103]. Norms are also helpful in estimating rates of convergence of iterative methods for solving linear and nonlinear equations and optimizing functions. Functional analysis, which deals with infinite-dimensional vector spaces, uses norms on functions.

We have already met the Euclidean vector norm $\|x\|$ on R^n. For most purposes, this norm suffices. It shares with other norms the four properties:

(a) $\|x\| \geq 0$,

(b) $\|x\| = 0$ if and only if $x = \mathbf{0}$,

(c) $\|cx\| = |c| \cdot \|x\|$ for every real number c,

(d) $\|x + y\| \leq \|x\| + \|y\|$.

Property (d) is known as the triangle inequality. To prove it for the Euclidean norm, we note that the Cauchy-Schwarz inequality implies

$$\begin{aligned} \|x+y\|^2 &= \|x\|^2 + 2x^*y + \|y\|^2 \\ &\leq \|x\|^2 + 2\|x\|\|y\| + \|y\|^2 \\ &= (\|x\| + \|y\|)^2 . \end{aligned}$$

One immediate consequence of the triangle inequality is the further inequality

$$\big|\, \|x\| - \|y\| \,\big| \leq \|x - y\|.$$

Two other simple but helpful norms are

$$\begin{aligned} \|x\|_1 &= \sum_{i=1}^{n} |x_i| \\ \|x\|_\infty &= \max_{1 \leq i \leq n} |x_i|. \end{aligned}$$

Some of the properties of these norms are explored in the problems.

An $m \times n$ matrix $A = (a_{ij})$ can be viewed as a vector in R^{mn}. Accordingly, we define its Euclidean norm

$$\|A\|_E = \left(\sum_{i=1}^{m} \sum_{j=1}^{n} a_{ij}^2 \right)^{1/2} = \sqrt{\operatorname{tr}(AA^*)} = \sqrt{\operatorname{tr}(A^*A)},$$

where $\operatorname{tr}(\cdot)$ is the matrix trace function. Our reasons for writing $\|A\|_E$ rather than $\|A\|$ will soon be apparent. In the meanwhile, the Euclidean matrix norm satisfies the additional condition

(e) $\|AB\| \leq \|A\| \cdot \|B\|$

for any two compatible matrices $A = (a_{ij})$ and $B = (b_{ij})$. Property (e) is verified by invoking the Cauchy-Schwarz inequality in

$$\begin{aligned} \|AB\|_E^2 &= \sum_{i,j} \Big| \sum_k a_{ik} b_{kj} \Big|^2 \\ &\leq \sum_{i,j} \Big(\sum_k a_{ik}^2 \Big) \Big(\sum_l b_{lj}^2 \Big) \\ &= \Big(\sum_{i,k} a_{ik}^2 \Big) \Big(\sum_{l,j} b_{lj}^2 \Big) \qquad (2.1) \\ &= \|A\|_E^2 \|B\|_E^2 . \end{aligned}$$

The Euclidean norm does not satisfy the natural condition

(f) $\|I\| = 1$

for an identity matrix I. Indeed, an easy calculation shows that $\|I\|_E = \sqrt{n}$ when I is $n \times n$.

To meet all of the conditions (a) through (f), we need to turn to induced matrix norms. Let $\|\cdot\|$ denote both the Euclidean norm on R^m and the Euclidean norm on R^n. The induced matrix norm on $m \times n$ matrices is defined by

$$\begin{aligned} \|A\| &= \sup_{x \neq \mathbf{0}} \frac{\|Ax\|}{\|x\|} \\ &= \sup_{\|x\|=1} \|Ax\|. \end{aligned} \tag{2.2}$$

The question of whether the indicated supremum exists is settled by the inequalities

$$\|Ax\| \;\leq\; \sum_{i=1}^{n} |x_i| \cdot \|Ae_{(i)}\| \;\leq\; \left(\sum_{i=1}^{n} \|Ae_{(i)}\|\right) \|x\|,$$

where $x = \sum_{i=1}^{n} x_i e_{(i)}$ and $e_{(i)}$ is the unit vector whose entries are all 0 except for $e_{ii} = 1$. All of the defining properties of a matrix norm are trivial to check for definition (2.2). For instance, consider property (e):

$$\begin{aligned} \|AB\| &= \sup_{\|x\|=1} \|ABx\| \\ &\leq \|A\| \sup_{\|x\|=1} \|Bx\| \\ &= \|A\| \cdot \|B\|. \end{aligned}$$

Definition (2.2) also clearly entails the equality $\|I\| = 1$.

The next proposition determines the value of $\|A\|$. In the proposition, $\rho(M)$ denotes the absolute value of the dominant eigenvalue of the matrix M. This quantity is called the spectral radius of M.

Proposition 2.2.1 *If $A = (a_{ij})$ is an $m \times n$ matrix, then*

$$\|A\| \;=\; \sqrt{\rho(A^*A)} \;=\; \sqrt{\rho(AA^*)} \;=\; \|A^*\|.$$

When A is symmetric, $\|A\|$ reduces to $\rho(A)$. The norms $\|A\|$ and $\|A\|_E$ satisfy

$$\|A\| \;\leq\; \|A\|_E \;\leq\; \sqrt{n}\|A\|. \tag{2.3}$$

Finally, when A is a row or column vector, the matrix and vector norms of A coincide.

Proof: Choose an orthonormal basis of eigenvectors $u_{(1)}, \ldots, u_{(n)}$ for the symmetric matrix A^*A with corresponding eigenvalues arranged so that $0 \leq \lambda_1 \leq \cdots \leq \lambda_n$. If $x = \sum_{i=1}^n c_i u_{(i)}$ is a unit vector, then $\sum_{i=1}^n c_i^2 = 1$, and

$$\begin{aligned} \|A\|^2 &= \sup_{\|x\|=1} x^* A^* A x \\ &= \sup_{\|x\|=1} \sum_{i=1}^n \lambda_i c_i^2 \\ &\leq \lambda_n. \end{aligned}$$

Equality is achieved when $c_n = \pm 1$ and all other $c_i = 0$. If A is symmetric with eigenvalues μ_i arranged so that $|\mu_1| \leq \cdots \leq |\mu_n|$, then the $u_{(i)}$ can be chosen to be the corresponding eigenvectors. In this case, clearly $\lambda_i = \mu_i^2$.

To prove that $\rho(A^*A) = \rho(AA^*)$, choose an eigenvalue $\lambda \neq 0$ of A^*A with corresponding eigenvector v. Multiplying the equation $A^*Av = \lambda v$ on the left by A produces $(AA^*)Av = \lambda Av$. Because $A^*Av = \lambda v$, the vector $Av \neq \mathbf{0}$. Thus, λ is an eigenvalue of AA^* with eigenvector Av. Likewise, any eigenvalue $\omega \neq 0$ of AA^* is also an eigenvalue of A^*A.

To verify the left bound of the pair of bounds (2.3), apply inequality (2.1) with $B = x$ in the definition of $\|A\|$. The right bound follows from

$$\textstyle\sum_{i=1}^m a_{ij}^2 \quad = \quad \|Ae_{(j)}\|^2 \quad \leq \quad \|A\|^2$$

by summing on j. Finally, suppose that A is a column vector. The two bounds (2.3) with $n = 1$ show that $\|A\| = \|A\|_E$. If A is a row vector, the same reasoning applied to A^* gives $\|A\| = \|A^*\| = \|A^*\|_E = \|A\|_E$. ■

2.3 Convergence and Completeness

A sequence $x_{(m)} \in \mathsf{R}^n$ converges to x, written $\lim_{m\to\infty} x_{(m)} = x$, provided $\lim_{m\to\infty} \|x_{(m)} - x\| = 0$. For convergence of $x_{(m)}$ to x to occur, it is necessary and sufficient that each component sequence x_{mi} converge to x_i. Convergence of a sequence of matrices is defined similarly using either the Euclidean norm $\|A\|_E$ or the induced matrix norm $\|A\|$. The pair of bounds (2.3) shows that the two norms are equivalent in testing convergence.

Convergent sequences of vectors or matrices enjoy many useful properties. Some of these are mentioned in the next proposition.

Proposition 2.3.1 *In the following list, once a limit is assumed to exist for an item, it is assumed to exist for all subsequent items. With this proviso, we have:*

(a) *If* $\lim_{m\to\infty} x_{(m)} = x$, *then* $\lim_{m\to\infty} \|x_{(m)}\| = \|x\|$.

(b) *If* $\lim_{m\to\infty} y_{(m)} = y$*, then*

$$\lim_{m\to\infty} x_{(m)}^* y_{(m)} = x^* y.$$

(c) *If* a *and* b *are real scalars, then*

$$\lim_{m\to\infty} \left[a x_{(m)} + b y_{(m)} \right] = a x + b y.$$

(d) *If* $\lim_{m\to\infty} M_{(m)} = M$ *for a sequence of matrices compatible with* x*, then*

$$\lim_{m\to\infty} M_{(m)} x_{(m)} = M x.$$

(e) *If* M *is square and invertible, then* $M_{(m)}^{-1}$ *exists for large* m *and*

$$\lim_{m\to\infty} M_{(m)}^{-1} = M^{-1}.$$

(f) *Finally, if* $\lim_{m\to\infty} N_{(m)} = N$ *for a sequence of matrices compatible with* M*, then*

$$\lim_{m\to\infty} M_{(m)} N_{(m)} = M N.$$

Proof: As a sample proof, part (d) follows from the inequalities

$$\begin{aligned} \|M_{(m)} x_{(m)} - M x\| &\le \|M_{(m)} x_{(m)} - M_{(m)} x\| + \|M_{(m)} x - M x\| \\ &\le \|M_{(m)}\| \cdot \|x_{(m)} - x\| + \|M_{(m)} - M\| \cdot \|x\|. \end{aligned}$$

Part (e) will be proved after Example 2.3.3. ■

In some situations, we know that the members of a sequence become progressively closer together. A Cauchy sequence $x_{(m)}$ exhibits a strong form of this phenomenon; namely, for every $\epsilon > 0$, there is an m such that $\|x_{(p)} - x_{(q)}\| \le \epsilon$ for all $p, q \ge m$. The real line R is complete in the sense that every Cauchy sequence possesses a limit. The rational numbers are incomplete by contrast because a sequence of rationals can converge to an irrational. The completeness of R carries over to R^n. Indeed, if $x_{(m)}$ is a Cauchy sequence, then with the Euclidean norm we have

$$|x_{pi} - x_{qi}| \le \|x_{(p)} - x_{(q)}\|.$$

This shows that each component sequence is Cauchy and consequently possesses a limit x_i. The vector x with components x_i then furnishes a limit for the vector sequence $x_{(m)}$.

Example 2.3.1 *Existence of Suprema and Infima*

The completeness of the real line is equivalent to the existence of least upper bounds or suprema. Consider a nonempty set $S \subset \mathsf{R}$ that is bounded above. If the set is finite, then its least upper bound is just its largest element. If the set is infinite, we choose a and b such that the interval $[a, b]$ contains an element of S and b is an upper bound of S. We can generate $\sup S$ by a bisection strategy. Bisect $[a, b]$ into the two subintervals $[a, (a+b)/2]$ and $[(a+b)/2, b]$. Let $[a_1, b_1]$ denote the left subinterval if $(a+b)/2$ provides an upper bound. Otherwise, let $[a_1, b_1]$ denote the right subinterval. In either case, $[a_1, b_1]$ contains an element of S. Now bisect $[a_1, b_1]$ and generate a subinterval $[a_2, b_2]$ by the same criterion. If we continue bisecting and choosing a left or right subinterval ad infinitum, then we generate two Cauchy sequences a_i and b_i with common limit c. By the definition of the sequence b_i, c furnishes an upper bound of S. By the definition of the sequence a_i, no bound of S is smaller than c. Establishing the existence of the greatest lower bound $\inf S$ for S bounded below proceeds similarly. If S is unbounded above, then $\sup S = \infty$, and if it is unbounded below, then $\inf S = -\infty$. ■

Example 2.3.2 *Limit Superior and Limit Inferior*

For a real sequence x_n. we define the limit superior and limit inferior by

$$\begin{aligned} \limsup_{n\to\infty} x_n &= \inf_m \sup_{n\geq m} x_n = \lim_{m\to\infty} \sup_{n\geq m} x_n \\ \liminf_{n\to\infty} x_n &= \sup_m \inf_{n\geq m} x_n = \lim_{m\to\infty} \inf_{n\geq m} x_n. \end{aligned}$$

If $\sup_n x_n = \infty$, then $\limsup_{n\to\infty} x_n = \infty$, and if $\lim_{n\to\infty} x_n = -\infty$, then $\limsup_{n\to\infty} x_n = -\infty$. From these definitions, one can also deduce that

$$\limsup_{n\to\infty} -x_n = -\liminf_{n\to\infty} x_n \tag{2.4}$$

and that

$$\liminf_{n\to\infty} x_n \leq \limsup_{n\to\infty} x_n. \tag{2.5}$$

The sequence x_n has a limit if and only if equality prevails in inequality (2.5). In this situation, the common value of the limit superior and inferior furnishes the limit of x_n. ■

Example 2.3.3 *Series Expansion for a Matrix Inverse*

If a square matrix M has norm $\|M\| < 1$, then we can write

$$(I - M)^{-1} = \sum_{i=0}^{\infty} M^i.$$

To verify this claim, we first prove that the partial sums $S_{(j)} = \sum_{i=0}^{j} M^i$ form a Cauchy sequence. This fact is a consequence of the inequalities

$$\begin{aligned}
\|S_{(k)} - S_{(j)}\| &= \Big\| \sum_{i=j+1}^{k} M^i \Big\| \\
&\leq \sum_{i=j+1}^{k} \|M^i\| \\
&\leq \sum_{i=j+1}^{k} \|M\|^i
\end{aligned}$$

for $k \geq j$ and the assumption $\|M\| < 1$. If we let S represent the limit of the $S_{(j)}$, then part (f) of Proposition 2.3.1 implies that $(I - M)S_{(j)}$ converges to $(I - M)S$. But $(I - M)S_{(j)} = I - M^{j+1}$ also converges to I. Hence, $(I - M)S = I$, and this verifies the claim $S = (I - M)^{-1}$. ■

With this result under our belts, we now demonstrate part (e) of Proposition 2.3.1. Because $\|M^{-1}(M - M_{(m)})\| \leq \|M^{-1}\| \cdot \|M - M_{(m)}\|$, the matrix inverse $[I - M^{-1}(M - M_{(m)})]^{-1}$ exists for large m. Therefore, we can write the inverse of

$$\begin{aligned}
M_{(m)} &= M - (M - M_{(m)}) \\
&= M[I - M^{-1}(M - M_{(m)})]
\end{aligned}$$

as

$$M_{(m)}^{-1} = [I - M^{-1}(M - M_{(m)})]^{-1} M^{-1}.$$

The proof of convergence is completed by noting the bound

$$\begin{aligned}
\|M_{(m)}^{-1} - M^{-1}\| &= \Big\| \sum_{i=1}^{\infty} [M^{-1}(M - M_{(m)})]^i M^{-1} \Big\| \\
&\leq \sum_{i=1}^{\infty} \|M^{-1}\|^i \|M - M_{(m)}\|^i \|M^{-1}\| \\
&= \frac{\|M^{-1}\|^2 \|M - M_{(m)}\|}{1 - \|M^{-1}\| \cdot \|M - M_{(m)}\|},
\end{aligned}$$

applying in the process the matrix analog of part (a) of Proposition 2.3.1.

Example 2.3.4 *Matrix Exponential Function*

The exponential of a square matrix M is given by the series expansion

$$e^M = \sum_{i=0}^{\infty} \frac{1}{i!} M^i.$$

To prove the convergence of the series, it again suffices to show that the partial sums $S_{(j)} = \sum_{i=0}^{j} \frac{1}{i!} M^i$ form a Cauchy sequence. The bound

$$\|S_{(k)} - S_{(j)}\| = \Big\| \sum_{i=j+1}^{k} \frac{1}{i!} M^i \Big\| \leq \sum_{i=j+1}^{k} \frac{1}{i!} \|M\|^i$$

for $k \geq j$ is just what we need.

The matrix exponential function has many interesting properties. For example, the function $N(t) = e^{tM}$ solves the differential equation

$$N'(t) = MN(t)$$

subject to the initial condition $N(0) = I$. Here t is a real parameter, and we differentiate the matrix $N(t)$ entry by entry. In Example 3.4.1 of Chapter 3, we will prove that $N(t) = e^{tM}$ is the one and only solution. The law of exponents $e^{A+B} = e^A e^B$ for commuting matrices A and B is another interesting property of the matrix exponential function. One way of proving the law of exponents is to observe that $e^{t(A+B)}$ and $e^{tA}e^{tB}$ both solve the differential equation

$$N'(t) = (A+B)N(t)$$

subject to the initial condition $N(0) = I$. Since the solution to such an initial value problem is unique, the two solutions must coincide at $t = 1$. ■

2.4 The Topology of R^n

Mathematics involves a constant interplay between the abstract and the concrete. We now consider some qualitative features of sets in R^n that generalize to more abstract spaces. For instance, there is the matter of boundedness. A set $S \subset \mathsf{R}^n$ is said to be bounded if it is contained in some ball $B(\mathbf{0}, r) = \{x \in \mathsf{R}^n : \|x\| < r\}$ of radius r centered at the origin $\mathbf{0}$. This concept takes on added importance when it is combined with the notion of closedness. A closed set is closed under the formation of limits. Thus, $S \subset \mathsf{R}^n$ is closed if for every convergent sequence $x_{(m)}$ taken from S, we have $\lim_{m\to\infty} x_{(m)} \in S$ as well.

It takes time and effort to appreciate the ramifications of these ideas. A few of the most pertinent are noted in the next proposition.

Proposition 2.4.1 *The collection of closed sets satisfy the following:*

(a) *The whole space* R^n *is closed.*

(b) *The empty set $\emptyset$ is closed.*

(c) *The intersection $S = \cap_\alpha S_\alpha$ of an arbitrary number of closed sets S_α is closed.*

(d) *The union $S = \cup_\alpha S_\alpha$ of a finite number of closed sets S_α is closed.*

Proof: All of these are easy. For part (d), observe that for any convergent sequence $x_{(m)}$ taken from S, one of the sets S_α must contain an infinite subsequence $x_{(m_k)}$. The limit of this subsequence exists and falls in S_α. ■

Some examples of closed sets are closed intervals $(-\infty, a]$, $[a, b]$, and $[b, \infty)$; closed balls $\{y \in \mathsf{R}^n : \|y - x\| \leq r\}$; spheres $\{y \in \mathsf{R}^n : \|y - x\| = r\}$; hyperplanes $\{x \in \mathsf{R}^n : z^*x = c\}$; and closed halfspaces $\{x \in \mathsf{R}^n : z^*x \leq c\}$. A closed set S of R^n is complete in the sense that all Cauchy sequences from S possess limits in S.

Example 2.4.1 *Finitely Generated Convex Cones*

A set C is a convex cone provided $\alpha u + \beta v$ is in C whenever the vectors u and v are in C and the scalars α and β are nonnegative. A finitely generated convex cone can be written as

$$C = \Big\{\sum_{i=1}^m \alpha_i v_{(i)} : \alpha_i \geq 0,\ i = 1, \ldots, m\Big\}.$$

Demonstrating that C is a closed set is rather subtle. Consider a sequence $u_{(j)} = \sum_{i=1}^m \alpha_{ji} v_{(i)}$ in C converging to a point u. If the vectors $v_{(1)}, \ldots, v_{(m)}$ are linearly independent, then the coefficients α_{ji} are the unique coordinates of $u_{(j)}$ in the finite-dimensional subspace spanned by the $v_{(i)}$. To recover α_{ji}, we can use the Gram-Schmidt orthogonalization procedure of linear algebra [18] to construct a vector $w_{(i)}$ satisfying $w_{(i)}^* v_{(k)} = 1_{\{i=k\}}$. If we take the inner product of $u_{(j)}$ with $w_{(i)}$, then it is clear that

$$\lim_{j\to\infty} \alpha_{ji} = \lim_{j\to\infty} w_{(i)}^* u_{(j)} = \alpha_i$$

exists and is nonnegative. Therefore, the limit $u = \sum_{i=1}^m \alpha_i v_{(i)}$ lies in C.

If we relax the assumption that the vectors are linearly independent, we must resort to an inductive argument to prove that C is closed. The case $m = 1$ is true because a single vector $v_{(1)}$ is linearly independent. Assume that the claim holds for $m - 1$ vectors. If the vectors $v_{(1)}, \ldots, v_{(m)}$ are linearly independent, then we are done. If the vectors $v_{(1)}, \ldots, v_{(m)}$ are linearly dependent, then there exist scalars $\beta_1, \ldots, \beta_m$, not all 0, such that $\sum_{i=1}^m \beta_i v_{(i)} = \mathbf{0}$. Without loss of generality, we can assume that $\beta_i < 0$ for at least one index i. We can express any point $u \in C$ as

$$u = \sum_{i=1}^m \alpha_i v_{(i)} = \sum_{i=1}^m (\alpha_i + t\beta_i) v_{(i)}$$

for an arbitrary scalar t. If we increase t gradually from 0, then there is a first value at which $\alpha_j + t\beta_j = 0$ for some index j. This shows that C can be decomposed as the union

$$C = \bigcup_{j=1}^{m} \Big\{ \sum_{i \neq j} \gamma_i v_{(i)} : \gamma_i \geq 0,\ i \neq j \Big\}.$$

Each of the convex cones $\{\sum_{i \neq j} \gamma_i v_{(i)} : \gamma_i \geq 0,\ i \neq j\}$ is closed by the induction hypothesis. Since a finite union of closed sets is closed, C itself is closed. ■

The complement $S^c = \mathsf{R}^n \setminus S$ of a closed set S is called an open set. Every $x \in S^c$ is surrounded by a ball $B(x, r)$ completely contained in S^c. If this were not the case, then we could construct a sequence of points $x_{(m)}$ from S converging to x, contradicting the closedness of S. This fact is the first of several mentioned in the following list.

Proposition 2.4.2 *The collection of open sets satisfy the following:*

(a) *Every open set is a union of balls, and every union of balls is an open set.*

(b) *The whole space* R^n *is open.*

(c) *The empty set* $\emptyset$ *is open.*

(d) *The union* $S = \cup_\alpha S_\alpha$ *of an arbitrary number of open sets* S_α *is open.*

(e) *The intersection* $S = \cap_\alpha S_\alpha$ *of a finite number of open sets* S_α *is open.*

Proof: Again these are easy. Parts (d) and (e) are consequences of the set identities

$$\begin{aligned} (\cap_\alpha S_\alpha)^c &= \cup_\alpha S_\alpha^c \\ (\cup_\alpha S_\alpha)^c &= \cap_\alpha S_\alpha^c \end{aligned}$$

and parts (c) and (d) of Proposition 2.4.1. ■

Some examples of open sets are open intervals $(-\infty, a)$, (a, b), and (b, ∞); balls $\{y \in \mathsf{R}^n : \|y - x\| < r\}$, and open halfspaces $\{x \in \mathsf{R}^n : z^*x < c\}$. Any open set surrounding a point is called a neighborhood of the point. Some examples of sets that are neither closed nor open are the unbalanced intervals $(a, b]$ and $[a, b)$, the discrete set $V = \{n^{-1} : n = 1, 2, \ldots\}$, and the rational numbers. If we append the limit 0 to the set V, then it becomes closed.

A boundary point x of a set S is the limit of a sequence of points from S and also the limit of a different sequence of points from S^c. Closed sets contain all of their boundary points, and open sets contain none of their boundary points. The interior of S is the largest open set contained within S. The closure of S is the smallest closed set containing S. For instance, the boundary of the ball $B(x,r) = \{y \in \mathsf{R}^n : \|y-x\| < r\}$ is the sphere $S(x,r) = \{y \in \mathsf{R}^n : \|y-x\| = r\}$. The closure of $B(x,r)$ is the closed ball $C(x,r) = \{y \in \mathsf{R}^n : \|y-x\| \leq r\}$, and the interior of $C(x,r)$ is $B(x,r)$.

A closed bounded set is said to be compact. Finite intervals $[a,b]$ are typical compact sets. Compact sets can be defined in several equivalent ways. The most important of these is the Bolzano-Weierstrass characterization. In preparation for this result, let us define a multidimensional interval $[a,b]$ in R^n to be the Cartesian product

$$[a,b] \quad = \quad [a_1,b_1] \times \cdots \times [a_n,b_n]$$

of n one-dimensional intervals. We will only consider closed intervals. The diameter of $[a,b]$ is the greatest separation between any two of its points; this clearly reduces to the distance $\|a-b\|$ between its extreme corners.

Proposition 2.4.3 (Bolzano-Weierstrass) *A set $S \subset \mathsf{R}^n$ is compact if and only if every sequence $x_{(m)}$ in S has a convergent subsequence $x_{(m_i)}$ with limit in S.*

Proof: Suppose every sequence $x_{(m)}$ in S has a convergent subsequence $x_{(m_i)}$ with limit in S. If S is unbounded, then we can define a sequence $x_{(m)}$ with $\|x_{(m)}\| \geq m$. Clearly, this sequence has no convergent subsequence. If S is not closed, then there is a convergent sequence $x_{(m)}$ with limit x outside S. Clearly, no subsequence of $x_{(m)}$ can converge to a limit in S. Thus, the subsequence property implies compactness.

For the converse, let $x_{(m)}$ be a sequence in the compact set S. Because S is bounded, it is contained in a multidimensional interval $[a,b]$. If infinitely many of the $x_{(m)}$ coincide, then these can be used to construct a constant subsequence that trivially converges to a point of S. Otherwise, let T_0 denote the infinite set $\cup_{m=1}^{\infty}\{x_{(m)}\}$.

The rest of the proof adapts the bisection strategy of Example 2.3.1. The first stage of the bisection divides $[a,b]$ into 2^n subintervals of equal volume. Each of these subintervals can be written as $[a_{(1)},b_{(1)}]$, where $a_{1j} = a_j$ and $b_{1j} = (a_j+b_j)/2$ or $a_{1j} = (a_j+b_j)/2$ and $b_{1j} = b_j$. There is no harm in the fact that these subintervals overlap along their boundaries. It is only vital to observe that one of the subintervals contains an infinite subset $T_1 \subset T_0$. Let us choose such a subinterval and label it using the generic notation $[a_{(1)},b_{(1)}]$. We now repeat the process inductively on a selected subinterval $[a_{(i)},b_{(i)}] \subset [a_{(i-1)},b_{(i-1)}]$ and an infinite subset $T_i \subset T_{i-1}$. At stage $i+1$ we divide $[a_{(i)},b_{(i)}]$ into 2^n subintervals of equal volume. Each of these subintervals can be written as $[a_{(i+1)},b_{(i+1)}]$, where $a_{i+1,j} = a_{ij}$ and $b_{i+1,j} = (a_{ij}+b_{ij})/2$ or $a_{i+1,j} = (a_{ij}+b_{ij})/2$ and $b_{i+1,j} = b_{ij}$. One of these subintervals, which we label $[a_{(i+1)},b_{(i+1)}]$ for convenience, contains an infinite subset $T_{i+1} \subset T_i$.

We continue this process ad infinitum, in the process choosing $x_{(m_i)}$ from T_i so that $m_i > m_{i-1}$. Because $T_i \subset [a_{(i)}, b_{(i)}]$ and the diameter of $[a_{(i)}, b_{(i)}]$ tends to 0, the subsequence $x_{(m_i)}$ is Cauchy. By virtue of the completeness of R^n, this subsequence converges to some point x, which necessarily belongs to the closed set S. ■

In many instances it is natural to consider a subset S of R^n as a topological space in its own right. Notions of distance and convergence carry over immediately, but we must exercise some care in defining closed and open sets. In the relative topology, a subset $T \subset S$ is closed if and only if it can be represented as the intersection $T = S \cap C$ of S with a closed set C of R^n. If T is closed in S, then the obvious choice of C is the closure of T in R^n. Likewise, $T \subset S$ is open in the relative topology if and only if it can be represented as the intersection $T = S \cap O$ of S with an open set O of R^n. These two definitions are consistent with an open set being the relative complement of a closed set and vice versa. They are also consistent with the development of continuous functions sketched in the next section.

2.5 Continuous Functions

Continuous functions are the building blocks of mathematical analysis. Continuity is such an intuitive notion that ancient mathematicians did not even bother to define it. Proper recognition of continuity had to wait until differentiability was thoroughly explored. Our approach to continuity emphasizes convergent sequences. A function $f(x)$ from R^m to R^n is said to be continuous at y if $f(x_{(i)})$ converges to $f(y)$ for every sequence $x_{(i)}$ that converges to y. If the domain of $f(x)$ is a subset S of R^m, then the sequences $x_{(i)}$ and the point y are confined to S. Finally, $f(x)$ is said to be continuous if it is continuous at every point y of its domain.

The definition of continuity through convergent sequences tends to be simpler to apply than the competing ϵ and δ approach of calculus. We leave it to the reader to show that the two definitions are fully equivalent. Either definition has powerful consequences. For instance, it is clear that a vector-valued function is continuous if and only if each of its component functions is continuous. Before enumerating other less obvious consequences, it is helpful to forge a few tools for recognizing and constructing continuous functions. Fortunately, the collection of continuous functions is closed under many standard algebraic operations. Here are a few examples.

Proposition 2.5.1 *Given that the vector-valued functions $f(x)$ and $g(x)$ and matrix-valued function $M(x)$ and $N(x)$ are continuous and compatible whenever necessary, the following algebraic combinations are continuous:*

(a) *The norm $\|f(x)\|$.*

(b) *The inner product $f(x)^*g(x)$.*

(c) *The linear combination* $\alpha f(x) + \beta g(x)$ *for real scalars* α *and* β.

(d) *The matrix-vector product* $M(x)f(x)$.

(e) *The matrix inverse* $M^{-1}(x)$ *when* $M(x)$ *is square and invertible.*

(f) *The matrix product* $M(x)N(x)$.

(g) *The functional composition* $f \circ g(x) = f[g(x)]$.

Proof: Parts (a) through (f) are all immediate by-products of Proposition 2.3.1 and the definition of continuity. For part (g), suppose $x_{(i)}$ tends to x. Then $f(x_{(i)})$ tends to $f(x)$, and so $f \circ g(x_{(x_i)})$ tends to $f \circ g(x)$. ■

Example 2.5.1 *Rational Functions*

Because the coordinate functions x_i of $x \in \mathsf{R}^n$ are continuous, all polynomials in these functions are continuous as well. For example, the determinant of a square matrix is a continuous function of the entries of the matrix. Any rational function of the coordinate functions x_i of $x \in \mathsf{R}^n$ is continuous where its denominator does not vanish. Finally, any linear transformation of one vector space into another is continuous. ■

Example 2.5.2 *Distance to a Set*

The distance $d(x, S)$ from a point $x \in \mathsf{R}^n$ to a set S is defined by

$$d(x, S) = \inf_{z \in S} \|z - x\|.$$

To prove that the function $d(x, S)$ is continuous in x, take the infimum over $z \in S$ of both sides of the triangle inequality

$$\|z - x\| \leq \|z - y\| + \|y - x\|.$$

This demonstrates that $d(x, S) \leq d(y, S) + \|y - x\|$. Reversing the roles of x and y then leads to the inequality

$$|d(x, S) - d(y, S)| \leq \|y - x\|,$$

establishing continuity. ■

In generalizing continuity to more abstract topological spaces, the characterizations in the next proposition are crucial.

Proposition 2.5.2 *The following conditions are equivalent for a function* $f(x)$ *from* $T \subset \mathsf{R}^m$ *to* R^n*:*

(a) $f(x)$ *is continuous.*

(b) *The inverse image* $f^{-1}(S)$ *of every closed set* S *is closed.*

(c) *The inverse image* $f^{-1}(S)$ *of every open set* S *is open.*

Proof: To prove that (a) implies (b), suppose $x_{(i)}$ is a sequence in $f^{-1}(S)$ tending to $x \in T$. Then the conclusion $\lim_{i\to\infty} f(x_{(i)}) = f(x)$ identifies $f(x)$ as an element of the closed set S and therefore x as belonging to $f^{-1}(S)$. Conditions (b) and (c) are equivalent because of the relation

$$f^{-1}(S)^c = f^{-1}(S^c)$$

between inverse images and set complements. Finally, to prove that (c) entails (a), suppose that $\lim_{i\to\infty} x_{(i)} = x$. For any $\epsilon > 0$, the inverse image of the ball $B[f(x), \epsilon]$ is open by assumption. Consequently, there exists a neighborhood $T \cap B(x, \delta)$ mapped into $B[f(x), \epsilon]$. In other words,

$$\|f(x_{(i)}) - f(x)\| < \epsilon$$

whenever $\|x_{(i)} - x\| < \delta$, which is sufficient to validate continuity. ■

Example 2.5.3 *Continuity of* $\sqrt[m]{x}$

The root function $f(x) = \sqrt[m]{x}$ is the functional inverse of the power function $g(x) = x^m$. We have already noted that $g(x)$ is continuous. On the interval $(0, \infty)$, it is also strictly increasing and maps the open interval (a, b) onto the open interval (a^m, b^m). Put another way, $f^{-1}[(a, b)] = (a^m, b^m)$. (Here we implicitly invoke the intermediate value property proved in Proposition 2.6.1.) Because the inverse image of a union of open intervals is a union of open intervals, application of part (c) of Proposition 2.5.2 establishes the continuity of $f(x)$. ■

Example 2.5.4 *The Set of Positive Definite Matrices*

A symmetric $n \times n$ matrix $M = (m_{ij})$ can be viewed as a point in R^m for $m = \binom{n}{2} + n$. To demonstrate that the subset S of positive definite matrices is open in R^m, we invoke the classical test for positive definiteness using the determinants of the principal submatrices of M. The kth of these submatrices $M_{(k)}$ is the $k \times k$ upper left block of M. If M is positive definite, then we can show that $M_{(k)}$ is positive definite by taking a nontrivial $k \times 1$ vector $x_{(k)}$ and padding it with zeros to construct a nontrivial $n \times 1$ vector x. It is then clear that $x_{(k)}^* M_{(k)} x_{(k)} = x^* M x > 0$. Because $M_{(k)}$ is positive definite, its determinant $\det M_{(k)} > 0$. Conversely, if all of the $\det M_{(k)} > 0$, then M itself is positive definite [68].

Given this background, we write

$$S = \bigcap_{k=1}^{n} \{M : \det M_{(k)} > 0\}.$$

Because the functions $\det M_{(k)}$ are continuous in the entries of M, the inverse images $\{M : \det M_{(k)} > 0\}$ of the open set $(0, \infty)$ are open. Since a finite intersection of open sets is open, S itself is an open set. ■

As opposed to inverse images, the image of a closed (open) set under a continuous function need not be closed (open). However, continuous functions do preserve compactness.

Proposition 2.5.3 *Suppose the continuous function $f(x)$ maps the compact set $S \subset \mathsf{R}^m$ into R^n. Then the image $f(S)$ is compact.*

Proof: Let $f(x_{(i)})$ be a sequence in $f(S)$. Extract a convergent subsequence $x_{(i_j)}$ of $x_{(i)}$ with limit $y \in S$. Then the continuity of $f(x)$ compels $f(x_{(i_j)})$ to converge to $f(y)$. ■

We now come to one of the most important results in optimization theory.

Proposition 2.5.4 (Weierstrass) *Let $f(x)$ be a continuous real-valued function defined on a set S of R^n. If the set $T = \{x \in S : f(x) \geq f(y)\}$ is compact for some $y \in S$, then $f(x)$ attains its supremum on S. Similarly, if $T = \{x \in S : f(x) \leq f(y)\}$ is compact for some $y \in S$, then $f(x)$ attains its infimum on S. Both conclusions apply when S itself is compact.*

Proof: Consider the question of whether the function $f(x)$ attains its supremum $u = \sup_{x \in S} f(x)$. The set $f(T)$ is bounded by virtue of Proposition 2.5.3, and the supremum of $f(x)$ on T coincides with u. For every integer i choose a point $x_{(i)} \in T$ such that $f(x_{(i)}) \geq u - 1/i$. In view of the compactness of T, we can extract a convergent subsequence of $x_{(i)}$ with limit $z \in T$. The continuity of $f(x)$ along this subsequence then implies that $f(z) = u$. ■

Example 2.5.5 *Closest Point in a Set*

To prove that the distance $d(x, S)$ is achieved for some $z \in S$, we must assume that S is closed. In finding the closest point to x in S, choose any point $y \in S$. The set $T = S \cap \{z : \|z - x\| \leq \|y - x\|\}$ is both closed and bounded and therefore compact. Proposition 2.5.4 now informs us that the continuous function $z \mapsto \|z - x\|$ attains its infimum on S. ■

Example 2.5.6 *Equivalence of Norms*

Every norm $\|x\|^\dagger$ on R^n is equivalent to the Euclidean norm $\|x\|$ in the sense that there exist positive constants c_l and c_u such that the inequalities

$$c_l\|x\| \quad \leq \quad \|x\|^\dagger \quad \leq \quad c_u\|x\| \tag{2.6}$$

hold for all x. To prove the right inequality in (2.6), let $e_{(1)}, \ldots, e_{(n)}$ denote the standard basis. Then conditions (c) and (d) defining a norm indicate that $x = \sum_{i=1}^n x_i e_{(i)}$ satisfies

$$\begin{aligned}\|x\|^\dagger &\leq \sum_{i=1}^n |x_i| \cdot \|e_{(i)}\|^\dagger \\ &\leq \|x\| \sum_{i=1}^n \|e_{(i)}\|^\dagger.\end{aligned}$$

This proves the upper bound with $c_u = \sum_{i=1}^n \|e_{(i)}\|^\dagger$.

To establish the lower bound, we note that property (c) of a norm allows us to restrict attention to the sphere $S = \{x : \|x\| = 1\}$. Now the function $x \mapsto \|x\|^\dagger$ is uniformly continuous on R^n because

$$\begin{aligned} \big| \|x\|^\dagger - \|y\|^\dagger \big| &\leq \|x-y\|^\dagger \\ &\leq c_u\|x-y\| \end{aligned}$$

follows from the upper bound just demonstrated. Since the sphere S is compact, the continuous function $x \to \|x\|^\dagger$ attains its lower bound c_l on S. In view of property (b) defining a norm, $c_l > 0$. ■

A function $f(x)$ is said to be uniformly continuous on its domain S if for every $\epsilon > 0$ there exists a $\delta > 0$ such that $\|f(y) - f(x)\| < \epsilon$ whenever $\|y - x\| < \delta$. This sounds like ordinary continuity, but the chosen δ does not depend on the pivotal point $x \in S$. One of the virtues of a compact domain is that it forces uniform continuity.

Proposition 2.5.5 (Heine) *Every continuous function $f(x)$ from a compact set S of R^m into R^n is uniformly continuous.*

Proof: Suppose $f(x)$ fails to be uniformly continuous. Then for some $\epsilon > 0$, there exist sequences $x_{(i)}$ and $y_{(i)}$ from S such that $\lim_{i\to\infty} \|x_{(i)} - y_{(i)}\| = 0$ and $\|f(x_{(i)}) - f(y_{(i)})\| \geq \epsilon$. Since S is compact, we can extract a subsequence of $x_{(i)}$ that converges to a point $u \in S$. Along the corresponding subsequence of $y_{(i)}$ we can extract a subsubsequence that converges to a point $v \in S$. Substituting the constructed subsubsequences for $x_{(i)}$ and $y_{(i)}$ if necessary, we may assume that $x_{(i)}$ and $y_{(i)}$ both converge to the same limit $u = v$. The condition $\|f(x_{(i)}) - f(y_{(i)})\| \geq \epsilon$ now contradicts the continuity of $f(x)$ at u. ■

Example 2.5.7 *Rigid Motions*

Uniform continuity certainly appears in the absence of compactness. One spectacular example is a rigid motion. By this we mean a function $f(x)$ of R^n into itself with the property $\|f(y) - f(x)\| = \|y - x\|$ for every choice of x and y. We can better understand the rigid motion $f(x)$ by investigating the translated rigid motion $g(x) = f(x) - f(\mathbf{0})$ that maps the origin $\mathbf{0}$ into itself. Because $g(x)$ preserves distances, it also preserves inner products. This fact is evident from the equalities

$$\begin{aligned} \|y-x\|^2 &= \|y\|^2 - 2y^*x + \|x\|^2 \\ \|g(y) - g(x)\|^2 &= \|g(y)\|^2 - 2g(y)^*g(x) + \|g(x)\|^2 \\ \|g(y)\|^2 &= \|y\|^2 \\ \|g(x)\|^2 &= \|x\|^2. \end{aligned}$$

The inner product identity

$$g(y)^*g(x) = y^*x$$

is only possible if $g(y)$ is linear. To demonstrate this assertion, note that $g(x)$ maps the standard orthonormal basis $e_{(1)}, \ldots, e_{(n)}$ onto the orthonormal basis $g(e_{(1)}), \ldots, g(e_{(n)})$. Because

$$\begin{aligned} g(\alpha x + \beta y)^* g(e_{(i)}) &= (\alpha x + \beta y)^* e_{(i)} \\ &= \alpha x^* e_{(i)} + \beta y^* e_{(i)} \\ &= [\alpha g(x) + \beta g(y)]^* g(e_{(i)}) \end{aligned}$$

holds for all i, it follows that $g(\alpha x + \beta y) = \alpha g(x) + \beta g(y)$. In other words, $g(x)$ is linear. The linear transformations that preserve angles and distances are precisely the orthogonal transformations. Thus, the rigid motion $f(x)$ reduces to an orthogonal transformation Ux followed by the translation $f(\mathbf{0})$. Conversely, it is trivial to prove that every such transformation

$$f(x) = Ux + f(\mathbf{0})$$

is a rigid motion. ■

2.6 Connectedness

Roughly speaking, a set is disconnected if it can be split into two pieces sharing no boundary. A set is connected if it is not disconnected. One way of making this vague distinction precise is to consider a set S disconnected if there exists a real-valued continuous function $\phi(x)$ defined on S and having range $\{0, 1\}$. The nonempty subsets $A = \phi^{-1}(0)$ and $B = \phi^{-1}(1)$ then constitute the two disconnected pieces of S. According to part (b) of Proposition 2.5.2, both A and B are closed. Because one is the complement of the other, both are also open.

Arcwise connectedness is a variation on the theme of connectedness. A set is said to be arcwise connected if for any pair of points x and y of the set there is a continuous function $f(t)$ from the interval $[0, 1]$ into the set satisfying $f(0) = x$ and $f(1) = y$. We will see shortly that arcwise connectedness implies connectedness. On open sets, the two notions coincide.

Can we identify the connected subsets of the real line? Intuition suggests that the only connected subsets are intervals. Here a single point x is viewed as the interval $[x, x]$. Suppose S is a connected subset of R, and let a and b be two points of S. In order for S to be an interval, every point $c \in (a, b)$ should be in S. If S fails to contain an intermediate point c, then we can define a continuous function $\phi(x)$ disconnecting S by taking $\phi(x) = 0$ for $x < c$ and $\phi(x) = 1$ for $x > c$. Thus, every connected subset must be an interval.

To prove the converse, suppose a disconnecting function $\phi(x)$ lives on an interval. Select points a and b of the interval with $\phi(a) = 0$ and $\phi(b) = 1$. Without loss of generality we can take $a < b$. On $[a, b]$ we now carry out the bisection strategy of Example 2.3.1, selecting the right or left subinterval at each stage so that the values of $\phi(x)$ at the endpoints of the selected subinterval disagree. Eventually, bisection leads to a subinterval contradicting the uniform continuity of $\phi(x)$ on $[a, b]$. Indeed, there is a number δ such that $|\phi(y) - \phi(x)| < 1$ whenever $|y - x| < \delta$; at some stage, the length of the subinterval containing points with both values of $\phi(x)$ falls below δ.

This result is the first of four characterizing connected sets.

Proposition 2.6.1 *Connected subsets of* R^n *have the following properties:*

(a) *A subset of the real line is connected if and only if it is an interval.*

(b) *The image of a connected set under a continuous function is connected.*

(c) *The union* $S = \cup_\alpha S_\alpha$ *of an arbitrary collection of connected subsets is connected if one of the sets* S_β *has a nonempty intersection* $S_\beta \cap S_\alpha$ *with every other set* S_α.

(d) *Every arcwise connected set* S *is connected.*

Proof: To prove part (b) let $f(x)$ be a continuous map from a connected set $S \subset \mathsf{R}^m$ into R^n. If the image $f(S)$ is disconnected, then there is a continuous function $\phi(x)$ disconnecting it. The composition $\phi \circ f(x)$ is continuous by part (g) of Proposition 2.5.1 and serves to disconnect S, contradicting the connectedness of S. To prove (c) suppose that the continuous function $\phi(x)$ disconnects the union S. Then there exists $y \in S_{\alpha_1}$ and $z \in S_{\alpha_2}$ with $\phi(y) = 0$ and $\phi(z) = 1$. Choose $u \in S_\beta \cap S_{\alpha_1}$ and $v \in S_\beta \cap S_{\alpha_2}$. If $\phi(u) \neq \phi(v)$, then $\phi(x)$ disconnects S_β. If $\phi(u) = \phi(v)$, then $\phi(y) \neq \phi(u)$ or $\phi(z) \neq \phi(v)$. In the former case $\phi(x)$ disconnects S_{α_1}, and in the latter case $\phi(x)$ disconnects S_{α_2}. Finally, to prove part (d), suppose the arcwise connected set S fails to be connected. Then there exists a continuous disconnecting function $\phi(x)$ with $\phi(y) = 0$ and $\phi(z) = 1$. Let $f(t)$ be an arc in S connecting y and z. The continuous function $\phi \circ f(t)$ then serves to disconnect $[0, 1]$. ■

Example 2.6.1 *The Intermediate Value Property*

Consider a continuous function $f(x)$ from an interval $[a, b]$ to the real line. The intermediate value theorem asserts that the image $f([a, b])$ coincides with the interval $[\min f(x), \max f(x)]$. This theorem, which is a consequence of properties (a) and (b) of Proposition 2.6.1, has many applications. For example, suppose $g(x)$ is a continuous function from $[0, 1]$ into $[0, 1]$. If $f(x) = g(x) - x$, then it is obvious that $f(0) \geq 0$ and $f(1) \leq 0$. It follows that $f(x) = 0$ for some x. In other words, $g(x)$ has a fixed point satisfying $g(x) = x$. ■

Example 2.6.2 *Connectedness of Spheres*

The sphere $S(x,r)$ in R^n is the image of the continuous map $y \mapsto x+ry/\|y\|$ of the domain $T = \mathsf{R}^n \setminus \mathbf{0}$. Hence, to prove connectedness when $n > 1$, it suffices to prove that T is connected. To achieve this, we argue that T is arcwise connected. Consider two points u and v in T. If $\mathbf{0}$ does not lie on the line segment between u and v, then we can use the function $f(t) = u + t(v-u)$ to connect u and v. If $\mathbf{0}$ lies on the line segment, choose any w not on the line determined by u and v. Now the continuous function

$$f(t) = \begin{cases} u + 2t(w-u) & t \in [0, \frac{1}{2}] \\ w + (2t-1)(v-w) & t \in [\frac{1}{2}, 1] \end{cases}$$

connects u and v. The sphere $S(x,r)$ in R reduces to the two points $x-r$ and $x+r$ and is disconnected. ■

2.7 Uniform Convergence

Many delicate issues of analysis revolve around the question of whether a given property of a sequence of functions $f_n(x)$ is preserved under a passage to a limit. As a simple example, consider the sequence $f_n(x) = x^n$ of continuous functions defined on the unit interval $[0,1]$. It is clear that $f_n(x)$ converges pointwise to the discontinuous function

$$f(x) = \begin{cases} 0 & 0 \le x < 1 \\ 1 & x = 1\,. \end{cases}$$

The failure of $f(x)$ to be continuous suggests that an additional hypothesis must be imposed. The key hypothesis is uniform convergence. This requires for each $\epsilon > 0$ that there exists an integer m such that $|f_n(x) - f(x)| < \epsilon$ for all $n \ge m$ and all x. Here the adjective "uniform" refers to the assumption that the same m works for all x. Of course, m is allowed to depend on ϵ.

Proposition 2.7.1 *Suppose the sequence of continuous functions $f_n(x)$ maps a domain $D \subset \mathsf{R}^j$ into R^k. If $f_n(x)$ converges uniformly to $f(x)$ on D, then $f(x)$ is also continuous.*

Proof: Choose $y \in D$ and $\epsilon > 0$, and take m so that $\|f_n(x) - f(x)\| < \frac{\epsilon}{3}$ for all $n \ge m$ and x. By virtue of the continuity of $f_m(x)$, there is a $\delta > 0$ such that $\|f_m(x) - f_m(y)\| < \frac{\epsilon}{3}$ whenever $\|x - y\| < \delta$. Assuming that y is fixed and $\|x-y\| < \delta$, we have

$$\begin{aligned} \|f(x) - f(y)\| &\le \|f(x) - f_m(x)\| + \|f_m(x) - f_m(y)\| + \|f_m(y) - f(y)\| \\ &< \frac{\epsilon}{3} + \frac{\epsilon}{3} + \frac{\epsilon}{3} \\ &= \epsilon. \end{aligned}$$

This shows that $f(x)$ is continuous at y. ■

Example 2.7.1 *Weierstrass M-Test*

Suppose the entries $g_k(x)$ of a sequence of continuous functions satisfy $\|g_k(x)\| \le M_k$, where $\sum_{k=1}^\infty M_k < \infty$. Then Cauchy's criterion and Proposition 2.7.1 together imply that the partial sums $f_n(x) = \sum_{k=1}^n g_k(x)$ converge uniformly to the continuous function $f(x) = \sum_{k=1}^\infty g_k(x)$. ■

2.8 Problems

1. Show that

$$\|x\|_q \le \|x\|_p \tag{2.7}$$
$$\|x\|_p \le n^{\frac{1}{p}-\frac{1}{q}}\|x\|_q \tag{2.8}$$

   when p and q are chosen from $\{1, 2, \infty\}$ and $p < q$. Here $\|x\|_2$ is the Euclidean norm on R^n. These inequalities are sharp. Equality holds in (2.7) when $x = (1, 0, \ldots, 0)^*$, and equality holds in (2.8) when $x = (1, 1, \ldots, 1)^*$.

2. Show that $\|x\|^2 \le \|x\|_\infty \|x\|_1 \le \sqrt{n}\|x\|^2$ for any vector $x \in \mathsf{R}^n$.

3. Prove that $1 \le \|I\|$ and $\|M\|^{-1} \le \|M^{-1}\|$ for any matrix norm satisfying the defining properties (a) through (e) of Section 2.2.

4. Define $\|M\|_{\max} = \max_{i,j} |m_{ij}|$ for $M = (m_{ij})$. Show that this defines a vector norm but not a matrix norm on $n \times n$ matrices M.

5. Demonstrate properties (2.4) and (2.5) of the limit superior and limit inferior. Also check that the sequence x_n has a limit if and only if equality holds in inequality (2.5).

6. Let $l = \limsup_{n\to\infty} x_n$. Show that:

   (a) $l = -\infty$ if and only if $\lim_{n\to\infty} x_n = -\infty$.

   (b) $l = +\infty$ if and only if for every positive integer m and real r there exists an $n \ge m$ with $x_n > r$.

   (c) l is finite if and only if (a) for every $\epsilon > 0$ there is an m such that $n \ge m$ implies $x_n < l + \epsilon$ and (b) for every $\epsilon > 0$ and every m there is an $n \ge m$ such that $x_n > l - \epsilon$.

   Similar properties hold for the limit inferior.

7. Let $x_{(m)}$ be a convergent sequence in R^n with limit x. Prove that the sequence $s_{(m)} = (x_{(1)} + \cdots + x_{(m)})/m$ of arithmetic means converges to x.

8. Show that

$$\lim_{x\to\infty} p(x)e^{-x} = 0$$

for every polynomial $p(x)$.

9. Prove that the set of invertible matrices is open and that the sets of symmetric and orthogonal matrices are closed in R^{n^2}.

10. A square matrix is nilpotent if $A^k = \mathbf{0}$ for some positive integer k. If A and B are nilpotent, then show that $A+B$ need not be nilpotent. If we add the hypothesis that A and B commute, then show that $A+B$ is nilpotent. Use Example 2.3.3 to construct the inverses of the matrices $I+A$ and $I-A$ for A nilpotent [29].

11. Show that e^{-M} is the matrix inverse of e^M. A skew symmetric matrix M satisfies $M^* = -M$. Show that e^M is orthogonal when M is skew symmetric.

12. Demonstrate that the matrix exponential function $M \mapsto e^M$ is continuous. (Hint: Apply the Weierstrass M-test.)

13. Demonstrate that the function

$$f(x) = \begin{cases} \frac{x_1 x_2}{x_1^2 - x_2^2} & |x_1| \neq |x_2| \\ 0 & \text{otherwise} \end{cases}$$

on R^2 is discontinuous at $\mathbf{0}$.

14. Let $f(x)$ and $g(x)$ be real-valued continuous functions defined on the same domain. Prove that $\max\{f(x), g(x)\}$ and $\min\{f(x), g(x)\}$ are continuous functions. Prove that the function $f(x) = \max_i x_i$ is continuous on R^n.

15. Define the function

$$f(x) = \begin{cases} x & x \text{ is rational} \\ 1-x & x \text{ is irrational} \end{cases}$$

on $[0,1]$. At what points is $f(x)$ continuous? What is the image $f([0,1])$?

16. Give an example of a continuous function that does not map an open set to an open set. Give another example of a continuous function that does not map a closed set to a closed set.

17. Show that the set of $n \times n$ orthogonal matrices is compact. (Hint: Show that every orthogonal matrix O has norm $\|O\| = 1$.)

18. Let $f(x)$ be a continuous function from a compact set $S \subset \mathsf{R}^m$ into R^n. If $f(x)$ is one-to-one, then demonstrate that the inverse function $f^{-1}(y)$ is continuous from $f(S)$ to S.

19. Let $C = A \times B$ be the Cartesian product of two subsets $A \subset \mathsf{R}^m$ and $B \subset \mathsf{R}^n$. Prove that:

    (a) C is closed in R^{m+n} if both A and B are closed.

    (b) C is open in R^{m+n} if both A and B are open.

    (c) C is compact in R^{m+n} if both A and B are compact.

    (d) C is connected in R^{m+n} if both A and B are connected.

20. Prove the converse of each of the assertions in Problem 19.

21. Without appeal to Proposition 2.5.5, show that every polynomial on R is uniformly continuous on a compact interval $[a, b]$.

22. Let $f(x)$ be uniformly continuous on R and satisfy $f(0) = 0$. Demonstrate that there exists a nonnegative constant c such that

$$|f(x)| \leq 1 + c|x|$$

    for all x [29].

23. Suppose that $f(x)$ is continuous on $[0, \infty)$ and $\lim_{x \to \infty} f(x)$ exists and is finite. Prove that $f(x)$ is uniformly continuous on $[0, \infty)$ [29].

24. Show that a hyperplane $\{x : z^* x = c\}$ in R^n is connected but that its complement is disconnected.

25. If the real-valued function $f(x)$ on $[a, b]$ is continuous and one-to-one, then prove that $f(x)$ is either strictly increasing or strictly decreasing.

26. On what domains do the sequences of functions

    (a) $f_n(x) = (1 + x/n)^n$

    (b) $f_n(x) = nx/(1 + n^2x^2)$

    (c) $f_n(x) = n^x$

    (d) $f_n(x) = x^{-1} \sin(nx)$

    (e) $f_n(x) = xe^{-nx}$

    (f) $f_n(x) = x^{2n}/(1 + x^{2n})$

    converge [28]? On what domains do they converge uniformly?

27. Suppose $f(x)$ is a continuous function on R. Demonstrate that the sequence

$$f_n(x) = \frac{1}{n}\sum_{k=0}^{n-1} f\left(x+\frac{k}{n}\right)$$

converges uniformly to a continuous function on every finite interval $[a, b]$ [29].

3

Differentiation

3.1 Introduction

Differentiation and integration are the two pillars on which all of calculus rests. For real-valued functions of a real variable, all of the major issues surrounding differentiation were settled long ago. For multivariate differentiation, there are still some subtleties and snares. We adopt a definition of differentiability that avoids most of the pitfalls and makes differentiation of vectors and matrices relatively painless. In later chapters, this definition also improves the clarity of exposition.

The main theme of differentiation is the short-range approximation of curved functions by linear functions. A differential gives a recipe for carrying out such a linear approximation. Most linear approximations can be improved by adding more terms in a Taylor series expansion. Adding quadratic terms brings in second differentials. We will meet these in the next chapter after we have mastered first differentials. Our current treatment stresses theory and counterexamples rather than the nuts and bolts of differentiation.

3.2 Univariate Derivatives

A real-valued function $f(x)$ defined on an open interval (a, b) possesses a derivative $f'(c)$ at $c \in (a, b)$ provided the limit

$$\lim_{x \to c} \frac{f(x) - f(c)}{x - c} = f'(c) \tag{3.1}$$

exists. Taking a sequential view of convergence, this means that for every sequence x_m converging to c we must have

$$\lim_{m\to\infty} \frac{f(x_m) - f(c)}{x_m - c} = f'(c).$$

In calculus, we learn the following rules for computing derivatives:

Proposition 3.2.1 *If $f(x)$ and $g(x)$ are differentiable functions on (a,b), then*

$$\begin{aligned}
\Big[\alpha f(x) + \beta g(x)\Big]' &= \alpha f'(x) + \beta g'(x) \\
\Big[f(x)g(x)\Big]' &= f'(x)g(x) + f(x)g'(x) \\
\Big[\frac{1}{f(x)}\Big]' &= -\frac{f'(x)}{f(x)^2}.
\end{aligned}$$

In the last formula we must assume $f(x) \neq 0$.

Proof: We will prove these in a more general context later. ■

In addition to the sum, product, and quotient rules, there is the chain rule

$$[f \circ g(x)]' = f' \circ g(x) g'(x)$$

governing the composition of differentiable functions. These rules can be combined in an endless variety of ways to solve concrete problems.

The standard repertoire of differentiable functions includes the derivatives nx^{n-1} of the monomials x^n and, via the cited rules, the derivatives of all polynomials and rational functions. These functions are supplemented by special functions such as $\ln x$, e^x, $\sin x$, and $\cos x$. Virtually all of the special functions can be defined by power series or as the solutions of differential equations. For instance, the system of differential equations

$$\begin{aligned}
(\cos x)' &= -\sin x \\
(\sin x)' &= \cos x
\end{aligned}$$

with the initial conditions $\cos 0 = 1$ and $\sin 0 = 0$ determines these trigonometric functions. We will take most of these facts for granted except to add in the case of $\cos x$ and $\sin x$ that the solution of the defining system of differential equations involves a particular matrix exponential.

Two further remarks are worth making at this junction. First, differentiation is a purely local operation. Second, differentiability at a point implies continuity at the same point but not vice versa. The functions

$$f_n(x) = \begin{cases} x^n & x \text{ rational} \\ 0 & x \text{ irrational} \end{cases}$$

illustrate the local character of continuity and differentiability. For $n > 0$ the functions $f_n(x)$ are continuous at the point 0 but discontinuous everywhere else. In contrast, $f_1'(0)$ fails to exist while $f_n'(0) = 0$ for all $n \geq 2$. In this instance, we must resort directly to the definition (3.1) to evaluate derivatives.

We have already mentioned Fermat's result that $f'(x)$ must vanish at any extreme point. For example, suppose that c is a local maximum of $f(x)$ on (a, b). If $f'(c) > 0$, then choose $\epsilon > 0$ such that $f'(c) - \epsilon > 0$. This choice then entails

$$\begin{aligned} f(x) &> f(c) + [f'(c) - \epsilon](x - c) \\ &> f(c) \end{aligned}$$

for all $x > c$ with $x - c$ sufficiently small, contradicting the assumption that c is a local maximum. If $f'(c) < 0$, we reach a similar contradiction using nearby points on the left of c.

Fermat's principle has some surprising consequences. Among these is the mean value property.

Proposition 3.2.2 *Suppose $f(x)$ is continuous on $[a, b]$ and differentiable on (a, b). Then there exists a point $c \in (a, b)$ such that*

$$f(b) - f(a) = f'(c)(b - a).$$

As a consequence:

(a) *If $f'(x) \geq 0$ for all $x \in (a, b)$, then $f(x)$ is increasing.*

(b) *If $f'(x) = 0$ for all $x \in (a, b)$, then $f(x)$ is constant.*

(c) *If $f'(x) \leq 0$ for all $x \in (a, b)$, then $f(x)$ is decreasing.*

Proof: Consider the function

$$g(x) = f(b) - f(x) + \frac{f(b) - f(a)}{b - a}(x - b).$$

Clearly, $g(x)$ is also continuous on $[a, b]$ and differentiable on (a, b). Furthermore, $g(a) = g(b) = 0$. It follows that $g(x)$ attains either a maximum or a minimum at some $c \in (a, b)$. At this point, $g'(c) = 0$, which is equivalent to the mean value property. ■

Example 3.2.1 *A Trigonometric Identity*

The function $f(x) = \cos^2 x + \sin^2 x$ has derivative

$$f'(x) = -2\cos x \sin x + 2 \sin x \cos x = 0.$$

Therefore, $f(x) = f(0) = 1$ for all x. ■

3.3 The Gauge Integral

Much of calculus deals with the interplay between differentiation and integration. The antiquated term "antidifferentiation" emphasizes the fact that differentiation and integration are inverses of one another. We will take it for granted that readers are acquainted with the mechanics of integration.

The first successful effort to put integration on a rigorous basis was undertaken by Riemann. In the early 20th century, Lebesgue defined a more sophisticated integral that addresses many of the limitations of the Riemann integral. However, even Lebesgue's integral has its defects. In the past few decades, mathematicians have expanded the definition of integration on the real line to include a wider variety of functions. The new integral emerging from these investigations is called the gauge integral or generalized Riemann integral [3, 50, 94, 28, 114, 116, 123]. The gauge integral subsumes the Riemann integral, the Lebesgue integral, and the improper integrals met in traditional advanced calculus courses. In contrast to the Lebesgue integral, the integrands of the gauge integral are not necessarily absolutely integrable.

It would take us too far afield to develop the gauge integral. We will be content here with describing some of its most basic properties. One of the advantages of the gauge integral is that many theorems hold with fewer qualifications. The fundamental theorem of calculus is a case in point. The commonly stated version of the fundamental theorem concerns a differentiable function $f(x)$ on an interval $[a,b]$. As all students of calculus know, the integral $\int_a^b f'(x)dx$ exists and equals $f(b)-f(a)$. Although this is true for the gauge integral, it unfortunately fails for certain pathological functions for the Lebesgue integral. Another rule is the integration-by-parts formula

$$\int_a^b f'(x)g(x)dx + \int_a^b f(x)g'(x)dx \quad = \quad f(b)g(b) - f(a)g(a),$$

which holds if either $f'(x)g(x)$ or $f(x)g'(x)$ is gauge integrable. A third rule is the substitution formula

$$\int_a^b f'[g(x)]g'(x)dx \quad = \quad \int_{g(a)}^{g(b)} f'(x)dx \tag{3.2}$$

for an increasing, differentiable function $g(x)$ on $[a,b]$ and a differentiable function $f(x)$ on $[g(a),g(b)]$. If $g(x)$ is decreasing rather than increasing, then formula (3.2) is valid subject to the obvious modifications.

Besides these familiar rules of calculus, we have the further elementary properties:

$$\int_a^b [\alpha f(x) + \beta g(x)]dx \quad = \quad \alpha \int_a^b f(x)dx + \beta \int_a^b g(x)dx$$

$$\begin{aligned}
\int_a^c f(x)dx &= \int_a^b f(x)dx + \int_b^c f(x)dx, \quad a < b < c \\
\int_a^b f(x)dx &\leq \int_a^b g(x)dx, \quad f(x) \leq g(x) \text{ for all } x \\
\Big| \int_a^b f(x)dx \Big| &\leq \int_a^b |f(x)|dx \,.
\end{aligned}$$

All monotonic, bounded functions and all continuous functions are gauge integrable. At a deeper level, gauge variants of the monotone and dominated convergence theorems are true. These theorems permit the interchange of limits and integrals. Finally, the gauge integral extends to multiple dimensions, where a version of Fubini's theorem holds for evaluating multidimensional integrals via iterated integrals.

3.4 Higher-Order Derivatives

The derivative of a function may itself be differentiable. Indeed, it makes sense to speak of the kth-order derivative of a function $f(x)$ if $f(x)$ is sufficiently smooth. Traditionally, the second-order derivative is denoted $f''(x)$ and an arbitrary kth-order derivative by $f^{(k)}(x)$. The mean value property is the simplest of an increasingly complex hierarchy of local polynomial approximations to $f(x)$. The next proposition makes this clear and offers an explicit estimate of the error in a finite Taylor expansion of $f(x)$.

Proposition 3.4.1 *Suppose $f(y)$ has a derivative of order $k+1$ on an open interval around the point x. Then for all y in the interval, we have*

$$f(y) = f(x) + \sum_{j=1}^{k} \frac{1}{j!} f^{(j)}(x)(y-x)^j + R_k(y), \tag{3.3}$$

where the remainder

$$R_k(y) = \frac{(y-x)^{k+1}}{k!} \int_0^1 f^{(k+1)}[x + t(y-x)](1-t)^k dt.$$

If $|f^{(k+1)}(z)| \leq b$ for all z between x and y, then

$$|R_k(y)| \leq \frac{b|y-x|^{k+1}}{(k+1)!}. \tag{3.4}$$

Proof: When $k = 0$, the Taylor expansion (3.3) reads

$$f(y) = f(x) + (y-x) \int_0^1 f'[x + t(y-x)]dt$$

and follows from the fundamental theorem of calculus and the substitution rule (3.2). Induction and the integration-by-parts formula

$$\begin{aligned}
& \int_0^1 f^{(k)}[x+t(y-x)](1-t)^{k-1}dt \\
= & -\frac{1}{k}f^{(k)}[x+t(y-x)](1-t)^k\Big|_0^1 \\
& + \frac{y-x}{k}\int_0^1 f^{(k+1)}[x+t(y-x)](1-t)^k dt \\
= & \frac{1}{k}f^k(x) + \frac{y-x}{k}\int_0^1 f^{(k+1)}[x+t(y-x)](1-t)^k dt
\end{aligned}$$

now validate the general expansion (3.3). The error estimate follows directly from the bound and the integral

$$\int_0^1 (1-t)^k dt = \frac{1}{k+1}.$$

Example 3.4.1 *Differential Equations and the Matrix Exponential*

The derivative of a vector or matrix-valued function $f(x)$ with domain (a,b) is defined entry by entry. We have already met the matrix-valued differential equation $N'(t) = MN(t)$ with initial condition $N(0) = I$. To demonstrate that $N(t) = e^{tM}$ is a solution, consider the difference quotient

$$\begin{aligned}
\frac{e^{(t+s)M} - e^{tM}}{s} &= \frac{1}{s}\sum_{j=1}^{\infty}\frac{(t+s)^j - t^j}{j!}M^j \\
&= M\sum_{j=1}^{\infty}\frac{t^{j-1}}{(j-1)!}M^{j-1} + \sum_{j=1}^{\infty}\frac{(t+s)^j - t^j - jt^{j-1}s}{sj!}M^j \\
&= Me^{tM} + \sum_{j=1}^{\infty}\frac{(t+s)^j - t^j - jt^{j-1}s}{sj!}M^j.
\end{aligned}$$

We now apply the error estimate (3.4) for the first-order Taylor expansion of the function $f(t) = t^j$. If c bounds $|t+s|$ for s near 0, it follows that

$$\left|(t+s)^j - t^j - jt^{j-1}s\right| \leq j(j-1)c^{j-2}s^2/2$$

and that

$$\begin{aligned}
\left\|\sum_{j=1}^{\infty}\frac{(t+s)^j - t^j - jt^{j-1}s}{sj!}M^j\right\| &\leq \frac{|s|}{2}\sum_{j=2}^{\infty}\frac{j(j-1)c^{j-2}}{j!}\|M\|^j \\
&= \frac{|s|}{2}\|M\|^2 e^{c\|M\|}.
\end{aligned}$$

This is enough to show that

$$\lim_{s \to 0} \left\| \frac{e^{(t+s)M} - e^{tM}}{s} - Me^{tM} \right\| = 0.$$

One can demonstrate that e^{tM} is the unique solution of the differential equation $N'(t) = MN(t)$ subject to $N(0) = I$ by considering the matrix $P(t) = e^{-tM}N(t)$ using any solution $N(t)$. Because the product rule of differentiation pertains to matrix multiplication as well as to ordinary multiplication,

$$P'(t) = -Me^{-tM}N(t) + e^{-tM}MN(t) = \mathbf{0}.$$

By virtue of part (b) of Proposition 3.2.2, $P(t)$ is the constant matrix $P(0) = I$. If we take $N(t) = e^{tM}$, then this argument demonstrates that e^{-tM} is the matrix inverse of e^{tM}. If we take $N(t)$ to be an arbitrary solution of the differential equation, then multiplying both sides of $e^{-tM}N(t) = I$ on the left by e^{tM} implies that $N(t) = e^{tM}$ as claimed. ■

Example 3.4.2 *Matrix Logarithm*

Let M be a square matrix with $\|M\| < 1$. It is tempting to define the logarithm of $I - M$ by the series expansion

$$\ln(I - M) = -\sum_{k=1}^{\infty} \frac{M^k}{k}$$

valid for scalars. This definition does not settle the question of whether

$$e^{\ln(I-M)} = I - M. \tag{3.5}$$

The traditional approach to such issues relies on Jordan canonical forms [69]. Here we would like to sketch an analytic proof. Consider the matrix-valued functions

$$\begin{aligned} f(t) &= e^{\ln(I-tM)} \\ f_n(t) &= e^{-\sum_{k=1}^{n} t^k M^k/k} \\ &= e^{-tM}e^{-t^2M^2/2} \cdots e^{-t^nM^n/n} \end{aligned}$$

of the scalar t. It is clear that $f_n(t)$ converges uniformly to $f(t)$ on every interval $[0, 1+\delta)$ for $\delta > 0$ small enough. Furthermore, the product rule, the chain rule, and the law of exponents show that

$$\begin{aligned} f_n'(t) &= -(M + tM^2 + \cdots + t^{n-1}M^n)f_n(t) \\ &= -M(I - t^nM^n)(I - tM)^{-1}f_n(t). \end{aligned}$$

Because $I - t^n M^n$ tends to I, it follows that

$$\lim_{n\to\infty} f_n'(t) = -M(I-tM)^{-1}f(t)$$

uniformly on $[0, 1+\delta)$. This in turn implies

$$\begin{aligned} f(t) - f(0) &= \lim_{n\to\infty} [f_n(t) - f_n(0)] \\ &= \lim_{n\to\infty} \int_0^t f_n'(s)ds \\ &= -M \int_0^t (I - sM)^{-1} f(s)ds. \end{aligned}$$

Differentiating this equation with respect to t produces the differential equation

$$f'(t) = -M(I-tM)^{-1}f(t) \tag{3.6}$$

with initial condition $f(0) = I$. Clearly $f(t) = I - tM$ is one solution of the differential equation (3.6). In view of Problem 9, this solution is unique. Comparing the two formulas for $f(t)$ at the point $t = 1$ now gives the desired conclusion (3.5). ∎

3.5 Partial Derivatives

There are several possible ways to extend differentiation to real-valued functions on R^n. The most familiar is the partial derivative

$$d_i f(x) = \frac{\partial}{\partial x_i} f(x) = \lim_{t\to 0} \frac{f(x + te_{(i)}) - f(x)}{t},$$

where $e_{(i)}$ is one of the standard unit vectors spanning R^n. There is nothing sacred about the coordinate directions. The directional derivative along the direction v is

$$d_v f(x) = \lim_{t\to 0} \frac{f(x+tv) - f(x)}{t}.$$

If we confine $t \geq 0$ in this limit, then we have a forward derivative along v.

To illustrate these definitions, consider the function

$$f(x) = \sqrt{|x_1 x_2|}$$

on R^2. It is clear that both partial derivatives are 0 at the origin. Along a direction $v = (v_1, v_2)^*$ with neither $v_1 = 0$ nor $v_2 = 0$, consider the difference quotient

$$\frac{f(\mathbf{0} + tv) - f(\mathbf{0})}{t} = \frac{|t|\sqrt{|v_1 v_2|}}{t}.$$

This has limit $\sqrt{|v_1v_2|}$ as long as we restrict $t > 0$. Thus, the forward directional derivative exists, but the full directional derivative does not.

For another example, let

$$f(x) = \begin{cases} x_1 + x_2 & \text{if } x_1 = 0 \text{ or } x_2 = 0 \\ 1 & \text{otherwise.} \end{cases}$$

This function is clearly discontinuous at the origin of R^2, but the partial derivatives $d_1 f(\mathbf{0}) = 1$ and $d_2 f(\mathbf{0}) = 1$ are perfectly behaved there. These and similar anomalies suggest the need for a carefully structured theory of differentiability. Such a theory is presented in the next section.

Second and higher-order partial derivatives are defined in the obvious way. For typographical convenience, we will occasionally employ such abbreviations as

$$d_{ij}^2 f(x) = \frac{\partial^2}{\partial x_i \partial x_j} f(x).$$

Readers will doubtless recall from calculus the equality of mixed second partial derivatives. This property can fail. For example, suppose we define $f(x) = g(x_1)$, where $g(x_1)$ is nowhere differentiable. Then $d_2 f(x)$ and $d_{12}^2 f(x)$ are identically 0 while $d_1 f(x)$ does not even exist. The key to restoring harmony is to impose continuity in a neighborhood of the current point.

Proposition 3.5.1 *Suppose the real-valued function $f(y)$ on R^2 has partial derivatives $d_1 f(y)$, $d_2 f(y)$, and $d_{12}^2 f(y)$ on some open set. If $d_{12}^2 f(y)$ is continuous at a point x in the set, then $d_{21}^2 f(x)$ exists and*

$$\frac{\partial^2}{\partial x_2 \partial x_1} f(x) = d_{21}^2 f(x) = d_{12}^2 f(x) = \frac{\partial^2}{\partial x_1 \partial x_2} f(x). \tag{3.7}$$

This result extends in the obvious way to the equality of second mixed partials for functions defined on open subsets of R^n for $n > 2$.

Proof: Consider the first difference

$$g(t) = f(x_1 + u_1, t) - f(x_1, t)$$

and the second difference

$$\begin{aligned} \Delta^2 f(x_1, x_2)(u_1, u_2) &= g(x_2 + u_2) - g(x_2) \\ &= f(x_1 + u_1, x_2 + u_2) - f(x_1, x_2 + u_2) \\ &\quad - f(x_1 + u_1, x_2) + f(x_1, x_2). \end{aligned}$$

Applying the mean value theorem twice gives

$$\begin{aligned} \Delta^2 f(x_1, x_2)(u_1, u_2) &= u_2 g'(x_2 + \theta_2 u_2) \\ &= u_2[d_2 f(x_1 + u_1, x_2 + \theta_2 u_2) - d_2 f(x_1, x_2 + \theta_2 u_2)] \\ &= u_1 u_2 d_{12}^2 f(x_1 + \theta_1 u_1, x_2 + \theta_2 u_2) \end{aligned}$$

for θ_1 and θ_2 in $(0,1)$. In view of the continuity of $d_{12}^2 f(y)$ at x, it follows that

$$\lim_{\|u\|\to 0} \frac{\Delta^2 f(x_1,x_2)(u_1,u_2)}{u_1 u_2} = d_{12}^2 f(x_1,x_2)$$

regardless of how the limit is approached. This proves the existence of the iterated limit

$$\begin{aligned} \lim_{u_2\to 0}\lim_{u_1\to 0} \frac{\Delta^2 f(x_1,x_2)(u_1,u_2)}{u_1 u_2} &= \lim_{u_2\to 0} \frac{d_1 f(x_1,x_2+u_2) - d_1 f(x_1,x_2)}{u_2} \\ &= d_{21}^2 f(x_1,x_2) \end{aligned}$$

and simultaneously the equality of mixed partials. ■

3.6 Differentials

The question of when a real-valued function is differentiable is perplexing because of the variety of possible definitions. In choosing an appropriate definition, we are governed by several considerations. First, it should be consistent with the classical definition of differentiability on the real line. Second, continuity at a point should be a consequence of differentiability at the point. Third, all directional derivatives should exist. Fourth, the differential should vanish wherever the function attains a local maximum or minimum on the interior of its domain. Fifth, the standard rules for combining differentiable functions should apply. Sixth, the logical proofs of the rules should be as transparent as possible. Seventh, the extension to vector-valued functions should be painless. Eighth and finally, our geometric intuition should be enhanced.

We now present a definition conceived by Constantin Carathéodory [15] and expanded by recent authors [1, 8, 13, 78] that fulfills these conditions. A real-valued function $f(y)$ on an open set $S \subset \mathsf{R}^m$ is said to be differentiable at $x \in S$ if a function $s(y,x)$ exists for y near x satisfying

$$\begin{aligned} f(y) - f(x) &= s(y,x)(y-x) \\ \lim_{y\to x} s(y,x) &= s(x,x). \end{aligned} \tag{3.8}$$

The row vector $s(y,x)$ is called a slope function. We will see in a moment that its limit $s(x,x)$ defines the differential $df(x)$ of $f(y)$ at x.

The standard definition of differentiability due to Fréchet reads

$$f(y) - f(x) = df(x)(y-x) + o(\|y-x\|)$$

for y near the point x. The matrix $df(x)$ appearing here is again termed the differential of $f(y)$ at x. Fréchet's definition is less convenient than

Carathéodory's because the former invokes approximate equality rather than true equality. Observe that the error $[s(y,x) - df(x)](y-x)$ under Carathéodory's definition satisfies

$$\|[s(y,x) - df(x)](y-x)\| \quad \leq \quad \|s(y,x) - df(x)\| \cdot \|y-x\|,$$

which is $o(\|y-x\|)$ as y tends to x in view of the continuity of $s(y,x)$. Thus, Carathéodory's definition implies Fréchet's definition. The converse is also true, as readers are asked to demonstrate in Problem 12.

Carathéodory's definition (3.8) has some immediate consequences. For example, on the real line we can choose

$$s(y,x) \quad = \quad \frac{f(y) - f(x)}{y - x} \tag{3.9}$$

as the slope function. However, there are more profound consequences. Definition (3.8) obviously compels $f(y)$ to be continuous at x. If we send t to 0 in the equation

$$\frac{f(x+tv) - f(x)}{t} \quad = \quad s(x+tv, x)v, \tag{3.10}$$

then it is clear that the directional derivative $d_v f(x)$ exists and equals $s(x,x)v$. The special case $v = e_{(i)}$ shows that the ith component of $s(x,x)$ reduces to the partial derivative $d_i f(x)$. Since $s(x,x)$ and $df(x)$ agree component by component, they are equal, and, in general, we have the formula $d_v f(x) = df(x)v$ for the directional derivative.

Fermat's rule that the differential of a function vanishes at an interior maximum or minimum point is also trivial to check in this context. Suppose x affords a local minimum of $f(y)$. Then $f(x+tv) - f(x) = ts(x+tv, x)v \geq 0$ for all v and small $t > 0$. Taking limits in the identity (3.10) now yields the conclusion $df(x)v \geq 0$. The only way this can hold for all v is for $df(x) = \mathbf{0}$. If x occurs on the boundary of the domain of $f(y)$, then we can still glean useful information. For example, if $f(y)$ is differentiable on the closed interval $[c,d]$ and c provides a local minimum, then the condition $f'(c) \geq 0$ must hold.

The extension of the definition of differentiability to vector-valued functions is equally simple. Suppose $f(y)$ maps an open subset $S \subset \mathsf{R}^m$ into R^n. Then $f(y)$ is said to be differentiable at $x \in S$ if there exists an $m \times n$ matrix-valued function $s(y,x)$ continuous at x and satisfying equation (3.8) for y near x. The limit $\lim_{y \to x} s(y,x) = df(x)$ is again called the differential of $f(y)$ at x. The rows of the differential are the differentials of the component functions of $f(x)$. Thus, $f(y)$ is differentiable at x if and only if each of its components is differentiable at x.

In many cases it is easy to identify a slope function. For example, the real-valued function $f(y) = y^2$ has slope function

$$\frac{y^2 - x^2}{y - x} \quad = \quad y + x.$$

A linear transformation $f(y) = My$ is differentiable with slope function $s(y, x) = M$. The real-valued coordinate functions y_i of $y \in \mathsf{R}^n$ fall into this category. It is worth stressing that while differentials are uniquely determined, slope functions are not. The real-valued function $f(y) = y_1 y_2$ is typical. Indeed, the identities

$$\begin{aligned} y_1 y_2 - x_1 x_2 &= x_2(y_1 - x_1) + y_1(y_2 - x_2) \\ y_1 y_2 - x_1 x_2 &= y_2(y_1 - x_1) + x_1(y_2 - x_2) \end{aligned}$$

define two equally valid slope functions at x.

The rules for calculating differentials of algebraic combinations of differentiable functions flow easily from Carathéodory's definition.

Proposition 3.6.1 *If the two functions $f(x)$ and $g(x)$ map the open set $S \subset \mathsf{R}^m$ differentiably into R^n, then the following combinations are differentiable and have the displayed differentials:*

(a) $d[\alpha f(x) + \beta g(x)] = \alpha df(x) + \beta dg(x)$ *for all constants α and β.*

(b) $d[f(x)^* g(x)] = f(x)^* dg(x) + g(x)^* df(x)$.

(c) $d[f(x)^{-1}] = -f(x)^{-2} df(x)$ *when $n = 1$ and $f(x) \neq 0$.*

Proof: Let the slope functions for $f(x)$ and $g(x)$ at x be $s_f(y, x)$ and $s_g(y, x)$. Rule (a) follows by taking the limit of the slope function identified in the equality

$$\alpha f(y) + \beta g(y) - \alpha f(x) - \beta g(x) = [\alpha s_f(y, x) + \beta s_g(y, x)](y - x).$$

Rule (b) stems from the equality

$$\begin{aligned} & f(y)^* g(y) - f(x)^* g(x) \\ = {} & [f(y) - f(x)]^* g(y) + f(x)^* [g(y) - g(x)] \\ = {} & g(y)^* s_f(y, x)(y - x) + f(x)^* s_g(y, x)(y - x), \end{aligned}$$

and rule (c) stems from the equality

$$\begin{aligned} f(y)^{-1} - f(x)^{-1} &= -f(y)^{-1} f(x)^{-1} [f(y) - f(x)] \\ &= -f(y)^{-1} f(x)^{-1} s_f(y, x)(y - x). \end{aligned}$$

■

The chain rule also has an exceptionally straightforward proof.

Proposition 3.6.2 *Suppose $f(x)$ maps the open set $S \subset \mathsf{R}^k$ differentiably into R^m and $g(z)$ maps the open set $T \subset \mathsf{R}^m$ differentiably into R^n. If the image $f(S)$ is contained in T, then the composition $g \circ f(x)$ is differentiable with differential $dg \circ f(x)\, df(x)$.*

Proof: Let $s_f(y,x)$ be the slope function of $f(y)$ for y near x and $s_g(z,w)$ be the slope function of $g(z)$ for z near $w = f(x)$. The chain rule follows after taking the limit of the slope function identified in the equality

$$\begin{aligned} g \circ f(y) - g \circ f(x) &= s_g[f(y), f(x)][f(y) - f(x)] \\ &= s_g[f(y), f(x)]s_f(y,x)(y - x). \end{aligned}$$

■

Of course, these rules do not exhaust the techniques for finding differentials. Here is an example where we must fall back on the definition. See Problem 10 for a generalization.

Example 3.6.1 *Differential of* $f_+(x)^2$

Suppose $f(x)$ is a real-valued differentiable function. Define the function $f_+(x) = \max\{f(x), 0\}$. In general, $f_+(x)$ is not a differentiable function, but $g(x) = f_+(x)^2$ is. This is obvious on the open set $\{x : f(x) < 0\}$, where $dg(x) = \mathbf{0}$, and on the open set $\{x : f(x) > 0\}$, where

$$dg(x) = 2f(x)df(x).$$

The troublesome points are those with $f(x) = 0$. Near such a point we have $f(y) - 0 = s(y,x)(y - x)$, and

$$g(y) - 0 = f_+(y)s(y,x)(y - x).$$

It follows that $dg(x) = f_+(x)df(x) = \mathbf{0}$ when $f(x) = 0$. In general, all three special cases can be summarized by the same rule $dg(x) = 2f_+(x)df(x)$. ■

An obvious consequence of constructing slopes through formula (3.9) is the symmetry condition $s(y,x) = s(x,y)$. There is little harm in imposing this restriction on slope functions in higher dimensions. For the most part, it is straightforward to check that our results hold with the symmetry restriction in place. In proving (b) of Proposition 3.6.1, we identified the asymmetric slope function $f(x)^* s_g(y,x) + g(y)^* s_f(y,x)$. An equally valid slope function is $f(y)^* s_g(y,x) + g(x)^* s_f(y,x)$. Averaging these gives the symmetric slope function

$$s_{fg}(y,x) = \frac{1}{2}[f(y) + f(x)]^* s_g(y,x) + \frac{1}{2}[g(y) + g(x)]^* s_f(y,x), \quad (3.11)$$

assuming that $s_f(y,x)$ and $s_g(y,x)$ are symmetric. Similarly in Example 3.6.1, the slope function

$$s_{f_+^2}(y,x) = [f_+(y) + f_+(x)]\, s_f(y,x)$$

is symmetric.

3.7 Multivariate Mean Value Theorem

The mean value theorem is one of the most useful tools of the differential calculus. Here is a simple generalization to multiple dimensions.

Proposition 3.7.1 *Let the function $f(x)$ map an open subset S of R^n to R^m. If $f(x)$ is differentiable on a neighborhood of $x \in S$, then*

$$f(y) = f(x) + \int_0^1 df[x + t(y - x)]\, dt\, (y - x) \tag{3.12}$$

for y near x.

Proof: Integrating component by component, we need only consider the case $m = 1$. According to the chain rule stated in Proposition 3.6.2, the real-valued function $g(t) = f[x + t(y - x)]$ of the scalar t has differential $dg(t) = df[x+t(y-x)](y-x)$. Because differentials and derivatives coincide on the real line, equality (3.12) follows from the fundamental theorem of calculus applied to $g(t)$. ■

The notion of continuous differentiability is ambiguous. On the one hand, we could say that $f(y)$ is continuously differentiable around x if it possesses a slope function $s(y, z)$ that is jointly continuous in its two arguments. This implies the continuity of $df(y) = s(y, y)$ around x. On the other hand, continuous differentiability suggests that we postulate the continuity of $df(y)$ to start with. In this case, we have at our disposal equation (3.12) giving the slope function

$$s(y, z) = \int_0^1 df[z + t(y - z)]\, dt. \tag{3.13}$$

If $df(y)$ is continuous near x, then this choice of $s(y, z)$ is jointly continuous in its arguments. Hence, the two definitions of continuous differentiability coincide. Note that the particular slope function (3.13) has the advantage of being symmetric.

Example 3.7.1 *Characterization of Constant Functions*

If the differential $df(x)$ of a real-valued function $f(x)$ vanishes on an open connected set $S \subset \mathsf{R}^n$, then $f(x)$ is constant there. To establish this fact, let z be an arbitrary point of S and define $T = \{x \in S : f(x) = f(z)\}$. Given the continuity of $f(x)$, it is obvious that T is closed relative to S. To prove that $T = S$, it suffices to show that T is also open relative to S. Indeed, if T and $S \setminus T$ are both nonempty open sets, then they disconnect S. Now any point $x \in T$ is contained in a ball $B(x, r) \subset S$. For $y \in B(x, r)$, formula (3.12) and the vanishing of the differential show that $f(y) = f(x)$. Thus, T is open. ■

Example 3.7.2 *Failure of Proposition 3.2.2*

Consider the function $f(x) = (\cos x, \sin x)^*$ from R to R^2. The form of the mean value theorem stated in Proposition 3.2.2 fails because there is no $x \in (0, 2\pi)$ satisfying

$$\begin{aligned} \mathbf{0} &= f(2\pi) - f(0) \\ &= \begin{pmatrix} -\sin x \\ \cos x \end{pmatrix} (2\pi - 0). \end{aligned}$$

In this regard recall Example 3.2.1. ■

In spite of this counterexample, the bound

$$\begin{aligned} \|f(y) - f(x)\| &\leq \left\| \int_0^1 df[x + t(y - x)]\, dt \right\| \cdot \|y - x\| \\ &\leq \sup_{t \in [0,1]} \|df[x + t(y - x)]\| \cdot \|y - x\| \qquad (3.14) \end{aligned}$$

is often an adequate substitute for theoretical purposes. Validation of inequality (3.14) would seem to require a careful examination of the gauge integral. Rather than launch a full-scale investigation, let us give an elementary derivation of inequality (3.14) using some of the concepts undergirding the gauge integral.

The gauge integral is defined through gauge functions. A gauge function is nothing more than a positive function $\delta(x)$ defined on a finite interval $[a, b]$. In approximating an integral of a function $h(t)$ over $[a, b]$ by a finite Riemann sum, it is important to sample the function most heavily in those regions where it changes most rapidly. Now by a Riemann sum we mean a sum

$$\sum_{i=0}^{n-1} h(t_i)(r_{i+1} - r_i)$$

where the mesh points $r_0 < r_1 < \cdots < r_n$ form a partition of $[a, b]$ with $r_0 = a$ and $r_n = b$, and the tags t_i are chosen so that $t_i \in [r_i, r_{i+1}]$. If $\delta(t_i)$ measures the rapidity of change of $h(t)$ near t_i, then it makes sense to take $\delta(t_i)$ small in regions of rapid change and to force r_i and r_{i+1} to belong to the interval $(t_i - \delta(t_i), t_i + \delta(t_i))$. A tagged partition with this property is called a δ-fine partition. Exercise 16 asserts that every gauge function possesses a δ-fine partition. The proof of this fact relies on the completeness of the real line.

How do we put these abstract definitions to work? Let

$$\mu = \sup_{t \in [0,1]} \|df[x + t(y - x)]\|,$$

and consider the function $g(t) = f[x+t(y-x)]$ defined on the unit interval. For $\epsilon > 0$ and $v = x + t(y - x)$, the continuity of the slope function $s(u, v)$ for $f(v)$ makes it possible to choose a gauge $\delta(t)$ such that

$$\|s(u, v)\| \le \mu + \epsilon$$

whenever u satisfies $\|u - v\| < \delta(t)$. On this neighborhood of v, it follows that

$$\|f(u) - f(v)\| \le (\mu + \epsilon)\|u - v\|.$$

Now select a δ-fine partition with mesh points $r_0, \ldots, r_n$ and corresponding tags $t_0, \ldots, t_{n-1}$. Noting the obvious telescoping sum, we calculate

$$\begin{aligned}
\|f(y) - f(x)\| &= \|g(1) - g(0)\| \\
&= \left\| \sum_{i=0}^{n-1} \{g(r_{i+1}) - g(t_i) + g(t_i) - g(r_i)\} \right\| \\
&\le \sum_{i=0}^{n-1} \{\|g(r_{i+1}) - g(t_i)\| + \|g(t_i) - g(r_i)\|\} \\
&\le (\mu + \epsilon) \sum_{i=0}^{n-1} \{(r_{i+1} - t_i) + (t_i - r_i)\} \cdot \|y - x\| \\
&= (\mu + \epsilon)\|y - x\|.
\end{aligned}$$

Since $\epsilon > 0$ was arbitrary, this completes the proof. ■

3.8 Inverse and Implicit Function Theorems

Two of the harder theorems involving differentials are the inverse and implicit function theorems. The definition of differentials through slope functions tends to make the proofs easier to understand. The current proof of the inverse function theorem also features an interesting optimization argument [1].

Proposition 3.8.1 *Let $f(x)$ map an open set $U \subset \mathsf{R}^n$ into R^n. If $f(x)$ is continuously differentiable on U and the square matrix $df(x)$ is invertible at the point z, then there exist neighborhoods V of z and W of $f(z)$ such that the inverse function $g(y)$ satisfying $g \circ f(x) = x$ exists and maps W onto V. Furthermore, $g(x)$ is continuously differentiable with differential $dg(x) = df \circ g(x)^{-1}$.*

Proof: Let $f(x)$ have slope function $s(y, x)$. If $s(y, x)$ is invertible as a square matrix, then the relations

$$f(y) - f(x) = s(y, x)(y - x)$$

and

$$y - x = s(y, x)^{-1}[f(y) - f(x)]$$

are equivalent. Now suppose we know $f(x)$ has functional inverse $g(y)$. Exchanging $g(y)$ for y and $g(x)$ for x in the second relation above produces

$$g(y) - g(x) = s[g(y), g(x)]^{-1}(y - x), \tag{3.15}$$

and taking limits gives the claimed differential, provided $g(y)$ is continuous. To prove the continuity of $g(y)$ and therefore the joint continuity of $s[g(y), g(x)]^{-1}$, it suffices to show that $\|s[g(y), g(x)]^{-1}\|$ is locally bounded. Continuity in this circumstance is then a consequence of the bound

$$\|g(y) - g(x)\| \leq \|s[g(y), g(x)]^{-1}\| \cdot \|y - x\|$$

flowing from equation (3.15). Thus, the difficult part of the proof consists in proving that $g(y)$ exists.

In view of the continuous differentiability of $f(x)$, there is some neighborhood V of z such that $s(y, x)$ is invertible for all x and y in V. Furthermore, we can take V small enough so that the norm $\|s(y, x)^{-1}\|$ is bounded there. On V, the equality $f(y) - f(x) = s(y, x)(y - x)$ shows that $f(x)$ is one-to-one. Hence, all that remains is to show that we can shrink V so that $f(x)$ maps V onto an open subset W containing $f(z)$.

For some $r > 0$, the ball $B(z, r)$ of radius r centered at z is contained in V. The sphere $S(z, r) = \{x : \|x - z\| = r\}$ and the ball $B(z, r)$ are disjoint and must have disjoint images under $f(x)$ because $f(x)$ is one-to-one on V. In particular, $f(z)$ is not contained in $f[S(z, r)]$. The latter set is compact because $S(z, r)$ is compact and $f(x)$ is continuous. Let $d > 0$ be the distance from $f(z)$ to $f[S(z, r)]$.

We now define the set W mentioned in the statement of the proposition to be the ball $B[f(z), d/2]$ and show that W is contained in the image of $B(z, r)$ under $f(x)$. Take any $y \in W = B[f(z), d/2]$. The particular function $h(x) = \|y - f(x)\|^2$ is differentiable and attains its minimum on the closed ball $C(z, r) = \{x : \|x - z\| \leq r\}$. This minimum is strictly less than $(d/2)^2$ because z certainly performs this well. Furthermore, the minimum cannot be reached at a point $u \in S(z, r)$, for then

$$\|f(u) - f(z)\| \leq \|f(u) - y\| + \|y - f(z)\| < 2d/2,$$

contradicting the choice of d. Thus, $h(x)$ reaches its minimum at some point u in the open set $B(z, r)$. Fermat's principle requires that the differential

$$dh(x) = -2[y - f(x)]^* df(x) \tag{3.16}$$

vanish at u. Given the invertibility of $df(u)$, we therefore have $f(u) = y$.

Finally replace V by the open set $B(z, r) \cap f^{-1}\{B[f(z), d/2]\}$ contained within it. Our arguments have shown that $f(x)$ is one-to-one from V onto $W = B[f(z), d/2]$. This allows us to define the inverse function $g(x)$ from W onto V and completes the proof. ∎

Example 3.8.1 *Polar Coordinates*

Consider the transformation

$$\begin{aligned} r &= \|x\| \\ \theta &= \arctan(x_2/x_1) \end{aligned}$$

to polar coordinates in R^2. A brief calculation shows that this transformation has differential

$$\begin{pmatrix} \frac{\partial}{\partial x_1} r & \frac{\partial}{\partial x_2} r \\ \frac{\partial}{\partial x_1} \theta & \frac{\partial}{\partial x_2} \theta \end{pmatrix} = \begin{pmatrix} \cos\theta & \sin\theta \\ -\frac{1}{r}\sin\theta & \frac{1}{r}\cos\theta \end{pmatrix}.$$

Because the determinant of the differential is $1/r$, the transformation is locally invertible wherever $r \neq 0$. Excluding the half axis $\{x_1 \leq 0, x_2 = 0\}$, the polar transformation maps the plane one-to-one and onto the open set $(0, \infty) \times (-\pi, \pi)$. Here the inverse transformation reduces to

$$\begin{aligned} x_1 &= r\cos\theta \\ x_2 &= r\sin\theta \end{aligned}$$

with differential

$$\begin{pmatrix} \cos\theta & -r\sin\theta \\ \sin\theta & r\cos\theta \end{pmatrix} = \begin{pmatrix} \cos\theta & \sin\theta \\ -\frac{1}{r}\sin\theta & \frac{1}{r}\cos\theta \end{pmatrix}^{-1}.$$

■

We now turn to the implicit function theorem.

Proposition 3.8.2 *Let $f(x, y)$ map an open set $S \subset \mathsf{R}^{m+n}$ into R^m. Suppose that $f(a, b) = \mathbf{0}$ and that $f(x, y)$ is continuously differentiable on a neighborhood of $(a, b) \in S$. If we split the differential*

$$df(x, y) = [d_1 f(x, y), d_2 f(x, y)]$$

into an $m \times m$ block and an $m \times n$ block, and if $d_1 f(a, b)$ is invertible, then there exists a neighborhood U of $b \in \mathsf{R}^n$ and a continuously differentiable function $g(y)$ from U into R^m such that $f[g(y), y)] = \mathbf{0}$. Furthermore, $g(y)$ is unique and has differential

$$dg(y) = -d_1 f[g(y), y]^{-1} d_2 f[g(y), y].$$

Proof: The function

$$h(x, y) = \begin{pmatrix} f(x, y) \\ y \end{pmatrix}$$

from S to R^{m+n} is continuously differentiable with differential

$$dh(x, y) = \begin{pmatrix} d_1 f(x, y) & d_2 f(x, y) \\ \mathbf{0} & I_n \end{pmatrix}.$$

To apply the inverse function theorem to $h(x,y)$, we must check that $dh(a,b)$ is invertible. This is straightforward because

$$\begin{pmatrix} d_1 f(a,b) & d_2 f(a,b) \\ \mathbf{0} & I_n \end{pmatrix} \begin{pmatrix} u \\ v \end{pmatrix} = \mathbf{0}$$

can only occur if $v = \mathbf{0}$ and $d_1 f(a,b)u = \mathbf{0}$. In view of the invertibility of $d_1 f(a,b)$, the second of these equalities entails $u = \mathbf{0}$.

Given the invertibility of $dh(a,b)$, the inverse function theorem implies that $h(x,y)$ possesses a continuously differentiable inverse that maps an open set W containing $(\mathbf{0}, b)$ onto an open set V containing (a,b). The inverse function takes a point (z,y) into the point $[k(z,y), y]$. Consider the function $g(y) = k(\mathbf{0}, y)$ defined for $(\mathbf{0}, y) \in W$. Being open, W contains a ball of radius r around $(\mathbf{0}, b)$. There is no harm in restricting the domain of $g(y)$ to the ball $U = \{y \in \mathsf{R}^n : \|y - b\| < r\}$. On this domain, $g(y)$ is continuously differentiable and $f[g(y), y] = \mathbf{0}$. Because $h(x,y)$ is one-to-one, $g(y)$ is uniquely determined.

Finally, we can identify the differential of $g(y)$ by constructing a slope function. If we let $f(x,y)$ have slope function $[s_1(u,v,x,y), s_2(u,v,x,y)]$ at (x,y), then

$$\begin{aligned} \mathbf{0} &= f[g(v), v] - f[g(y), y] \\ &= s_1[g(v), v, g(y), y][g(v) - g(y)] + s_2[g(v), v, g(y), y](v - y). \end{aligned}$$

The invertibility of $s_1[g(v), v, g(y), y]$ for v near y therefore gives

$$g(v) - g(y) = -s_1[g(v), v, g(y), y]^{-1} s_2[g(v), v, g(y), y](v - y).$$

In the limit as v approaches y, we recover the differential $dg(y)$. ■

Example 3.8.2 *Circles and the Implicit Function Theorem*

The function $f(x,y) = x^2 + y^2 - r^2$ has differential $df(x,y) = (2x, 2y)$. Unless $a = 0$, the conditions of the implicit function theorem apply at (a,b), and

$$\begin{aligned} g(y) &= \operatorname{sgn}(a)\sqrt{r^2 - y^2} \\ g'(y) &= -\frac{y}{x}. \end{aligned}$$

Clearly, $g(y)$ satisfies $g(y)^2 + y^2 - r^2 = 0$. ■

3.9 Differentials of Matrix-Valued Functions

To find the differential of a matrix-valued function, we turn the matrix into a vector by sequentially stacking its columns, with the leftmost column on top and the rightmost column on the bottom [90]. This stacking operator is denoted by $\text{vec}(M)$ for a matrix M. Construction of matrices in this fashion is intimately connected to the matrix Kronecker product. Let $A = (a_{ij})$ be an $r \times s$ matrix and $B = (b_{ij})$ a $t \times u$ matrix. The Kronecker product $A \otimes B$ is the $rt \times su$ block matrix

$$A \otimes B = \begin{pmatrix} a_{11}B & \cdots & a_{1s}B \\ \vdots & \ddots & \vdots \\ a_{r1}B & \cdots & a_{rs}B \end{pmatrix}.$$

One of the most important properties of the Kronecker product is the easily checked multiplication rule

$$(A \otimes B)(C \otimes D) = (AC) \otimes (BD). \tag{3.17}$$

Problem 20 explores several other interesting properties of matrix Kronecker products.

To get a feel for the connection between the linear operator vec and the Kronecker product, readers can convince themselves of the identities

$$\begin{aligned} \text{vec}(uv^*) &= v \otimes u \\ \text{vec}(A)^* \,\text{vec}(B) &= \text{tr}(A^*B) \end{aligned} \tag{3.18}$$

involving two vectors u and v and two matrices A and B. Of more immediate relevance for our purposes is the identity

$$\text{vec}(ABC) = (C^* \otimes A)\,\text{vec}(B). \tag{3.19}$$

Two special cases are

$$\begin{aligned} \text{vec}(AB) &= (I \otimes A)\,\text{vec}(B) \\ \text{vec}(BC) &= (C^* \otimes I)\,\text{vec}(B) \end{aligned}$$

for appropriately dimensioned identity matrices.

To prove equality (3.19), assume that B has columns $b_{(1)}, \ldots, b_{(q)}$. The standard basis $e_{(1)}, \ldots, e_{(q)}$ of R^q permits us to write B as the sum of outer products

$$B = \sum_{i=1}^{q} b_{(i)} e_{(i)}^*.$$

Applying this representation and equations (3.17) and (3.18), we calculate

$$\text{vec}(ABC) = \sum_{i=1}^{q} \text{vec}(A b_{(i)} e_{(i)}^* C)$$

$$
\begin{aligned}
&= \sum_{i=1}^{q} \text{vec}(Ab_{(i)}[C^* e_{(i)}]^*) \\
&= \sum_{i=1}^{q} (C^* e_{(i)}) \otimes (Ab_{(i)}) \\
&= (C^* \otimes A) \sum_{i=1}^{q} e_{(i)} \otimes b_{(i)} \\
&= (C^* \otimes A) \sum_{i=1}^{q} \text{vec}(b_{(i)} e_{(i)}^*) \\
&= (C^* \otimes A) \, \text{vec}(B).
\end{aligned}
$$

Because of space limitations, we will rest content with presenting a few examples of how these ideas play out in practice. Suppose that $M(x)$ is a differentiable matrix of the vector x. This implies the existence of a slope function $s_M(y, x)$ satisfying

$$
\text{vec}\, M(y) - \text{vec}\, M(x) = s_M(y, x)(y - x).
$$

To prove that $M(x)^{-1}$ is differentiable, we must find a suitable slope function. We do this by stacking columns in the identity

$$
M(y)^{-1} - M(x)^{-1} = -M(y)^{-1}[M(y) - M(x)]M(x)^{-1}.
$$

This gives

$$
\begin{aligned}
& \text{vec}\, M(y)^{-1} - \text{vec}\, M(x)^{-1} \\
= & -\text{vec}\left\{ M(y)^{-1}[M(y) - M(x)]M(x)^{-1} \right\} \\
= & -[M(x)^{-1}]^* \otimes M(y)^{-1}[\text{vec}\, M(y) - \text{vec}\, M(x)] \\
= & -[M(x)^{-1}]^* \otimes M(y)^{-1} s_M(y, x)(y - x).
\end{aligned}
$$

In view of the fact that $M(y)^{-1}$ is continuous in y, we have achieved our goal. If we insist on a symmetric slope function, then we start with the identity

$$
M(y)^{-1} - M(x)^{-1} = -M(x)^{-1}[M(y) - M(x)]M(y)^{-1}
$$

and wind up with the slope function $-[M(y)^{-1}]^* \otimes M(x)^{-1} s_M(y, x)$. Averaging this with the previous slope yields a symmetric slope.

If we undo these Kronecker product differentials, the result is sometimes interpreted as

$$
d\left[M(x)^{-1}\right] = -M(x)^{-1} dM(x) M(x)^{-1}, \tag{3.20}
$$

but this equation is ambiguous unless we take $dM(x)^{-1}$ and $dM(x)$ to mean infinitesimal changes in $M(x)^{-1}$ and $M(x)$ brought about by an infinitesimal change in x. Alternatively, we view equation (3.20) as signifying

$$\frac{\partial}{\partial x_i}\left[M(x)^{-1}\right] = -M(x)^{-1}\frac{\partial}{\partial x_i}M(x)M(x)^{-1}$$

for each i.

Sometimes it is simpler to calculate the old-fashioned way with partial derivatives. For example, consider $\det M(x)$. In this case, we exploit the determinant expansion

$$\det M = \sum_j m_{ij}M_{ij} \tag{3.21}$$

of a square matrix M in terms of the entries m_{ij} and corresponding cofactors M_{ij} of its ith row. If we ignore the dependence of $M(x)$ on x and view M exclusively as a function of its entries m_{ij}, then the expansion (3.21) gives

$$\frac{\partial}{\partial m_{ij}}\det M = M_{ij}.$$

The chain rule therefore implies

$$\begin{aligned}\frac{\partial}{\partial x_i}\det M(x) &= \sum_{jk}\frac{\partial}{\partial m_{jk}}\det M(x)\frac{\partial}{\partial x_i}m_{jk}(x)\\ &= \sum_{jk}M_{jk}(x)\frac{\partial}{\partial x_i}m_{jk}(x).\end{aligned}$$

This can be simplified by noting that the matrix with entry $(\det M)^{-1}M_{ji}$ in row i and column j is M^{-1}. It follows that

$$\frac{\partial}{\partial x_i}\det M(x) = \det M(x)\operatorname{tr}\left[M(x)^{-1}\frac{\partial}{\partial x_i}M(x)\right]$$

when $M(x)$ is invertible. If $\det M(x)$ is positive, for instance if $M(x)$ is positive definite, then we have the even cleaner formula

$$\frac{\partial}{\partial x_i}\ln\det M(x) = \operatorname{tr}\left[M(x)^{-1}\frac{\partial}{\partial x_i}M(x)\right].$$

This is sometimes expressed in the ambiguous but suggestive form

$$d\left[\ln\det M(x)\right] = \operatorname{tr}\left[M(x)^{-1}dM(x)\right].$$

The transformation of a matrix-valued function into a vector-valued function is occasionally an unnecessary distraction. Consider the matrix-valued function $f(M) = M^2$ for $n \times n$ matrices. If we adopt Fréchet's definition of a differential, it is natural to expand

$$\begin{aligned} f(M+N) - f(M) &= (M+N)^2 - M^2 \\ &= NM + MN + N^2 \\ &= NM + MN + o(\|N\|), \end{aligned} \tag{3.22}$$

and identify the linear map $N \mapsto df(M)(N) = NM + MN$ as the differential of $f(M)$. The error bound $o(\|N\|)$ in the expansion (3.22) relies on the inequality $\|N^2\| \le \|N\|^2$. This differentiation strategy extends readily to matrix-valued functions defined by convergent power series $\sum_{n=0}^{\infty} c_n M^n$.

3.10 Problems

1. For each positive integer n and real number x, find the derivative, if possible, of the function

$$f_n(x) = \begin{cases} x^n \sin\left(\frac{1}{x}\right) & x \neq 0 \\ 0 & x = 0 \,. \end{cases}$$

   Pay particular attention to the point 0.

2. Show that the function

$$f(x) = x \ln(1 + x^{-1})$$

   is strictly increasing on $(0, \infty)$ and satisfies $\lim_{x \to 0} f(x) = 0$ and $\lim_{x \to \infty} f(x) = 1$ [29].

3. Let $h(x) = f(x)g(x)$ be the product of two functions that are each k times differentiable. Derive Leibnitz's formula

$$h^{(k)}(x) = \sum_{j=0}^{k} \binom{k}{j} f^{(j)}(x) g^{(k-j)}(x).$$

4. Assume that the real-valued functions $f(y)$ and $g(y)$ are differentiable at the real point x. If (a) $f(x) = g(x) = 0$, (b) $g(y) \neq 0$ for y near x, and (c) $g'(x) \neq 0$, then demonstrate L'Hôpital's rule

$$\lim_{y \to x} \frac{f(y)}{g(y)} = \frac{f'(x)}{g'(x)}.$$

5. Suppose the positive function $f(x)$ satisfies the inequality

$$f'(x) \le c f(x)$$

   for some constant $c \ge 0$ and all $x \ge 0$. Prove that $f(x) \le e^{cx} f(0)$ for $x \ge 0$ [29].

6. Let $f(x)$ be a differentiable curve mapping $[a, b]$ into R^n. Show that $\|f(x)\|$ is constant if and only if $f(x)$ and $f'(x)$ are orthogonal for all x.

7. Suppose the constants $c_0, \ldots, c_k$ satisfy the condition

$$c_0 + \frac{c_1}{2} + \cdots \frac{c_k}{k+1} = 0.$$

Demonstrate that the polynomial $p(x) = c_0 + c_1 x + \cdots + c_k x^k$ has a root on the interval $[0, 1]$.

8. Let $f(x)$ be a real-valued function on R. A fixed point of $f(x)$ satisfies $x = f(x)$. Prove the following assertions:

   (a) If $|f'(x)| < 1$ for all x, then $f(x)$ has at most one fixed point.

   (b) If $|f'(x)| \leq c < 1$ for all x, then $f(x)$ has a fixed point that is the limit of any iteration sequence $x_k = f(x_{k-1})$.

   (c) The function $f(x) = x + (1 + e^x)^{-1}$ satisfies $|f'(x)| < 1$ for all x but lacks a fixed point.

9. Consider the ordinary differential equation $M'(t) = N(t)M(t)$ with initial condition $M(0) = A$ for $n \times n$ matrices. If A is invertible, then demonstrate that any two solutions coincide in a neighborhood of 0. (Hint: If $P(t)$ and $Q(t)$ are two solutions, then differentiate the product $P(t)^{-1}Q(t)$ and apply equation (3.20).)

10. Let $f(x)$ and $g(x)$ be real-valued functions defined on a neighborhood of y in R^n. If $f(x)$ is differentiable at y and has $f(y) = 0$, and $g(x)$ is continuous at y, then prove that $f(x)g(x)$ is differentiable at y.

11. Suppose the real-valued functions $f_1(s), \ldots, f_n(x)$ are each differentiable at the point y in their common open domain S. If we let $f(x) = \max\{f_1(x), \ldots, f_n(x)\}$ and J be the set of indices i satisfying $f_i(y) = f(y)$, then prove that the forward directional derivative of $f(x)$ exists for each direction v and equals $\max_{i \in J}\{df_i(y)v\}$. The example $|x| = \max\{-x, x\}$ shows that the full directional derivative does not exist in general.

12. Demonstrate that Fréchet's definition implies Carathéodory's definition of differentiability [1, 8]. (Hint: Suppose that $f(y)$ is Fréchet differentiable at x. Define the slope function

$$s(y, x) = \frac{1}{\|y - x\|^2}[f(y) - f(x) - df(x)(y - x)](y - x)^* + df(x)$$

for $y \neq x$, and check that it tends to $df(x)$ as y tends to x.)

13. A real-valued function $f(y)$ is said to be Gâteaux differentiable at x if there exists a vector g such that

$$\lim_{t \downarrow 0} \frac{f(x+tv) - f(x)}{t} = g^* v$$

for all unit vectors v. In other words, all directional derivatives exist and depend linearly on the direction v. In general, Gâteaux differentiability does not imply differentiability. However, if $f(y)$ satisfies a Lipschitz condition $|f(y) - f(z)| \leq c\|y - z\|$ in a neighborhood of x, then prove that Gâteaux differentiability at x implies differentiability at x with $g = \nabla f(x)$. In Chapter 5, the proof of Proposition 5.3.1 shows that a convex function is locally Lipschitz around each of its interior points. Thus, Gâteaux differentiability and differentiability are equivalent for a convex function at an interior point of its domain.

14. On the set $\{x \neq \mathbf{0}\}$, demonstrate that $\|x\|$ is differentiable with differential $d\|x\| = \|x\|^{-1}x^*$.

15. Suppose the vector-valued function $x(t)$ is differentiable in the scalar t. Show that $\|x(t)\|' \leq \|x'(t)\|$ whenever $x(t) \neq \mathbf{0}$.

16. Prove that a gauge function $\delta(t)$ on a finite interval $[a, b]$ possesses a δ-fine partition. (Hint: Let d be the supremum of the set of $c \in [a, b]$ such that $[a, c]$ possesses a δ-fine partition.)

17. Continuing Problem 1, show that $f_4(x)$ is continuously differentiable on R, that its image is an open subset of R, and yet $f(x)$ is not one-to-one on any interval around 0. What bearing does this have on the inverse function theorem?

18. Demonstrate that the function

$$f(x) = \begin{cases} x + 2x^2 \sin\left(\frac{1}{x}\right) & x \neq 0 \\ 0 & x = 0 \end{cases}$$

has a bounded derivative $f'(x)$ for all x and $f'(0) \neq 0$, yet $f(x)$ is not one-to-one on any interval around 0. Why does the inverse function theorem fail in this case?

19. Consider the equation $f(x) = tg(x)$ determined by the continuously differentiable functions $f(x)$ and $g(x)$ from R into R. If $f(0) = 0$ and $f'(0) \neq 0$, then show that in a suitably small interval $|t| < \delta$ there is a unique continuously differentiable function $x(t)$ solving the equation and satisfying $x(0) = 0$. Prove that $x'(0) = g(0)/f'(0)$.

20. Demonstrate the following facts about the Kronecker product of two matrices:

(a) $\alpha(A \otimes B) = (\alpha A) \otimes B = A \otimes (\alpha B)$ for any scalar α.

(b) $(A \otimes B)^* = A^* \otimes B^*$.

(c) $(A + B) \otimes C = A \otimes C + B \otimes C$.

(d) $A \otimes (B + C) = A \otimes B + A \otimes C$.

(e) $(A \otimes B) \otimes C = A \otimes (B \otimes C)$.

(f) $(A \otimes B)(C \otimes D) = (AC) \otimes (BD)$.

(g) If A and B are invertible square matrices, then

$$(A \otimes B)^{-1} = A^{-1} \otimes B^{-1}.$$

(h) If λ is an eigenvalue of the square matrix A with algebraic multiplicity r and μ is an eigenvalue of the square matrix B with algebraic multiplicity s, then $\lambda\mu$ is an eigenvalue of $A \otimes B$ with algebraic multiplicity rs.

(i) If A and B are square matrices, $\operatorname{tr}(A \otimes B) = \operatorname{tr}(A)\operatorname{tr}(B)$.

(j) If A is an $m \times m$ matrix, and B is an $n \times n$ matrix, then

$$\det(A \otimes B) = \det(A)^n \det(B)^m.$$

All asserted operations involve matrices of compatible dimensions. (Hint: For part (h), let $A = USU^{-1}$ and $B = VTV^{-1}$ be the Jordan canonical forms of A and B. Check that $S \otimes T$ is upper triangular.)

21. Suppose the matrix-valued function $M(x)$ is differentiable. Prove that $\operatorname{tr}[M(x)^*M(x)]$ is differentiable and find its differential. Argue that the differential has infinitesimal representation $2\operatorname{tr}[M(x)^*dM(x)]$.

22. Let $M(x)$ and $N(x)$ be two matrix-valued, differentiable functions such that $M(x)N(x)$ makes sense. Prove that $M(x)N(x)$ is differentiable and find its differential. Argue that the differential has infinitesimal representation $dM(x)N(x) + M(x)dN(x)$.

23. Suppose that the differential $df(x)$ of the continuously differentiable function $f(x)$ from R^m to R^n has full rank at y. Show that $f(x)$ is one-to-one in a neighborhood of y. Note that $m \le n$ must hold.

24. Consider the function $f(M) = M + M^2$ defined on $n \times n$ matrices. Show that the range of $f(M)$ contains a neighborhood of the trivial matrix $\mathbf{0}$ [29]. (Hint: Compute the differential of $f(M)$ and apply the inverse function theorem.)

25. Calculate the differentials of the matrix-valued functions e^M and $\ln(I - M)$. What are the values of these differentials at $M = \mathbf{0}$? (Hint: See the final paragraph of Section 3.9.)

4

Karush-Kuhn-Tucker Theory

4.1 Introduction

In the current chapter, we study the problem of minimizing a real-valued function $f(x)$ subject to the constraints

$$\begin{aligned} g_i(x) &= 0, \qquad 1 \leq i \leq p \\ h_j(x) &\leq 0, \qquad 1 \leq j \leq q. \end{aligned}$$

All of these functions share some open set $U \subset \mathsf{R}^n$ as their domain. Of course, maximizing $f(x)$ is equivalent to minimizing $-f(x)$, so there is no loss of generality in considering minimization. The function $f(x)$ is called the objective function, the functions $g_i(x)$ are called equality constraints, and the functions $h_j(x)$ are called inequality constraints. Any point $x \in U$ satisfying all of the constraints is said to be feasible. A constraint $h_j(x)$ is active at the feasible point x provided $h_j(x) = 0$; it is inactive if $h_j(x) < 0$. In general, we will assume that the feasible region is nonempty.

In exploring solutions to the above constrained minimization problem, we will meet a generalization of the Lagrange multiplier rule fashioned independently by Karush and Kuhn and Tucker. Under fairly weak regularity conditions, the rule holds at all extrema. In contrast to this necessary condition, sufficient conditions for an extremum involve second derivatives. To state and prove the most useful sufficient condition, we must confront second differentials and what it means for a function to be twice differentiable. The matter is straightforward conceptually but computationally messy. Fortunately, we can build on the material presented in Chapter 3.

4.2 The Multiplier Rule

Before embarking on the long and interesting proof of the multiplier rule, we turn to linear programming as a specific example of constrained optimization. A huge literature has grown up around this single application.

Example 4.2.1 *Linear Programming*

If the objective function $f(x)$ and the constraints $g_i(x)$ and $h_j(x)$ are all affine functions $z^*x + c$, then the constrained minimization problem is termed linear programming. In the literature on linear programming, the standard linear program is posed as one of minimizing $f(x) = z^*x$ subject to the linear equality constraints $Vx = d$ and the nonnegativity constraints $x_i \geq 0$ for all $1 \leq i \leq n$. The inequality constraints are abbreviated as $x \geq \mathbf{0}$. To show that the standard linear program encompasses our apparently more general version of linear programming, we note first that we can omit the affine constant in the objective function $f(x)$. The p linear equality constraints

$$0 \quad = \quad g_i(x) \quad = \quad v_{(i)}^* x - d_i$$

are already in the form $Vx = d$ if we define V to be the $p \times n$ matrix with ith row $v_{(i)}^*$ and d to be the $p \times 1$ vector with ith entry d_i. The inequality constraint $h_j(x) \leq 0$ can be elevated to an equality constraint $h_j(x) + y_j = 0$ by introducing an additional variable y_j called a slack variable with the stipulation that $y_j \geq 0$. If any of the variables x_i is not already constrained by $x_i \geq 0$, then we can introduce what are termed free variables $u_i \geq 0$ and $w_i \geq 0$ so that $x_i = u_i - w_i$ and replace x_i everywhere by this difference. ■

In proving the multiplier rule, it turns out that one must restrict the behavior of the constraints at a local extremum to avoid redundant constraints. There is more than one way of achieving this. We will impose a constraint qualification.

Definition 4.2.1 *Mangasarian-Fromovitz Constraint Qualification*

This condition [91] holds at a feasible point x provided the differentials $dg_i(x)$ are linearly independent and there exists a vector v with $dg_i(x)v = 0$ for all i and $dh_j(x)v < 0$ for all inequality constraints $h_j(x)$ active at x. The vector v is a tangent vector in the sense that infinitesimal motion from x along v stays within the feasible region. ■

Because the Mangasarian-Fromovitz condition is difficult to check, we will consider the simpler sufficient condition of Kuhn and Tucker [77] in the next section. In the meantime, we state and prove the Lagrange multiplier rule extended to inequality constraints by Karush and Kuhn and Tucker. Our proof reproduces McShane's lovely argument, which substitutes penalties for constraints [95].

Proposition 4.2.1 *Suppose the objective function $f(y)$ of the constrained optimization problem has a local minimum at the feasible point x. If $f(y)$ and the various constraint functions are continuously differentiable near x, then there exists a unit vector of Lagrange multipliers $\lambda_0, \ldots, \lambda_p$ and $\mu_1, \ldots, \mu_q$ such that*

$$\lambda_0 \nabla f(x) + \sum_{i=1}^{p} \lambda_i \nabla g_i(x) + \sum_{j=1}^{q} \mu_j \nabla h_j(x) \quad = \quad \mathbf{0}. \tag{4.1}$$

Moreover, each of the multipliers λ_0 and μ_j is nonnegative, and $\mu_j = 0$ whenever $h_j(x) < 0$. If the constraint functions satisfy the constraint qualification at x, then we can take $\lambda_0 = 1$.

Proof: Without loss of generality, we assume $x = \mathbf{0}$ and $f(\mathbf{0}) = 0$. By renumbering the inequality constraints if necessary, we also suppose that the first r of them are active at $\mathbf{0}$ and the last $q - r$ of them are inactive at $\mathbf{0}$. Now choose $\delta > 0$ so that (a) the closed ball

$$C(\mathbf{0}, \delta) \quad = \quad \{y \in \mathsf{R}^n : \|y\| \leq \delta\}$$

is contained in the open domain U, (b) $\mathbf{0}$ is the minimum point of $f(y)$ in $C(\mathbf{0}, \delta)$ subject to the constraints, (c) the objective and constraint functions are continuously differentiable in $C(\mathbf{0}, \delta)$, and (d) the constraints $h_j(x)$ inactive at $\mathbf{0}$ are inactive throughout $C(\mathbf{0}, \delta)$.

On the road to our ultimate goal, consider the functions

$$h_{j+}(y) \quad = \quad \max\{h_j(y), 0\}.$$

Using these functions, we now prove that for each $0 < \epsilon \leq \delta$, there exists an $\alpha > 0$ such that

$$f(y) + \|y\|^2 + \alpha \sum_{i=1}^{p} g_i(y)^2 + \alpha \sum_{j=1}^{r} h_{j+}(y)^2 \quad > \quad 0 \tag{4.2}$$

for all y with $\|y\| = \epsilon$. This is not an entirely trivial claim to prove because $f(y)$ can be negative on $C(\mathbf{0}, \delta)$ outside the feasible region.

Suppose the claim is false. Then there is a sequence of points $y_{(m)}$ with $\|y_{(m)}\| = \epsilon$ and a sequence of numbers α_m tending to ∞ such that

$$\begin{aligned} & f(y_{(m)}) + \|y_{(m)}\|^2 \\ \leq \quad & -\alpha_m \sum_{i=1}^{p} g_i(y_{(m)})^2 - \alpha_m \sum_{j=1}^{r} h_{j+}(y_{(m)})^2. \end{aligned} \tag{4.3}$$

Because the closed ball $C(\mathbf{0}, \epsilon)$ is compact, the sequence $y_{(m)}$ has a convergent subsequence, which without loss of generality we take to be the sequence itself. Denoting the limit of the sequence by z, clearly $\|z\| = \epsilon$. Dividing both sides of inequality (4.3) by $-\alpha_m$ and sending m to ∞ produce

$$\sum_{i=1}^{p} g_i(z)^2 + \sum_{j=1}^{r} h_{j+}(z)^2 = 0. \tag{4.4}$$

It follows that z is feasible with $f(z) \geq f(\mathbf{0}) = 0$. However, inequality (4.3) requires that each $f(y_{(m)}) \leq -\epsilon^2$. Since this last relation is preserved in the limit, we reach a contradiction and consequently establish the validity of inequality (4.2).

Our next goal is to prove that there exists a point u and a unit vector $(\lambda_0, \lambda_1, \ldots, \lambda_p, \mu_1, \ldots, \mu_r)^*$ such that (a) $\|u\| < \epsilon$, (b) each of the multipliers λ_0 and μ_j is nonnegative, and (c)

$$\lambda_0[\nabla f(u) + 2u] + \sum_{i=1}^{p} \lambda_i \nabla g_i(u) + \sum_{j=1}^{r} \mu_j \nabla h_j(u) = \mathbf{0}. \tag{4.5}$$

Observe here that the distinction between active and inactive constraints comes into play again. To prove the Lagrange multiplier rule (4.5), define

$$F(y) = f(y) + \|y\|^2 + \alpha \sum_{i=1}^{p} g_i(y)^2 + \alpha \sum_{j=1}^{r} h_{j+}(y)^2$$

using the α satisfying condition (4.2).

Given that $F(y)$ is continuous, there is a point u giving the unconstrained minimum of $F(y)$ on the compact set $C(\mathbf{0}, \epsilon)$. Because this point satisfies $F(u) \leq F(\mathbf{0}) = 0$, it is impossible that $\|u\| = \epsilon$ in view of inequality (4.2). Thus, u falls in the interior of $C(\mathbf{0}, \epsilon)$ where $\nabla F(u) = \mathbf{0}$ must occur. The gradient condition $\nabla F(u) = \mathbf{0}$ can be expressed as

$$\nabla f(u) + 2u + \alpha \sum_{i=1}^{p} 2g_i(u)\nabla g_i(u) + \alpha \sum_{j=1}^{r} 2h_{j+}(u)\nabla h_j(u) = \mathbf{0}, \tag{4.6}$$

invoking the differentiability of the functions $h_{j+}(y)^2$ derived in Example 3.6.1 of Chapter 3. If we divide equality (4.6) by the norm of the vector

$$v = [1, 2\alpha g_1(u), \ldots, 2\alpha g_p(u), 2\alpha h_{1+}(u), \ldots, 2\alpha h_{r+}(u)]^*$$

and redefine the Lagrange multipliers accordingly, then the multiplier rule (4.5) holds with each of the multipliers λ_0 and μ_j nonnegative.

Now choose a sequence $\epsilon_m > 0$ tending to 0 and corresponding points $u_{(m)}$ where the Lagrange multiplier rule (4.5) holds. The sequence of unit vectors $(\lambda_{m0}, \ldots, \lambda_{mp}, \mu_{m1}, \ldots, \mu_{mr})^*$ has a convergent subsequence with limit $(\lambda_0, \ldots, \lambda_p, \mu_1, \ldots, \mu_r)^*$ that is also a unit vector. Replacing the sequence $u_{(m)}$ by the corresponding subsequence $u_{(m_k)}$ allows us to take limits along $u_{(m_k)}$ in equality (4.5) and achieve equality (4.1). Observe that $u_{(m_k)}$ converges to $\mathbf{0}$ because $\|u_{(m_k)}\| \leq \epsilon_{m_k}$.

Finally, suppose the constraint qualification holds at the local minimum $\mathbf{0}$ and that $\lambda_0 = 0$. If all of the nonnegative multipliers μ_j are 0, then at least one of the λ_i with $1 \leq i \leq p$ is not 0. But this contradicts the linear independence of the $dg_i(\mathbf{0})$. Now consider the vector v guaranteed by the constraint qualification. Taking its inner product with both sides of equation (4.1) gives

$$\sum_{j=1}^{r} \mu_j dh_j(\mathbf{0})v = 0,$$

contradicting the assumption that $dh_j(\mathbf{0})v < 0$ for all $1 \leq j \leq r$ and the fact that at least one $\mu_j > 0$. Thus, $\lambda_0 > 0$, and we can divide equation (4.1) by λ_0. ∎

Example 4.2.2 *Application to an Inequality*

Let us demonstrate the inequality

$$\frac{x_1^2 + x_2^2}{4} \leq e^{x_1+x_2-2}$$

subject to the constraints $x_1 \geq 0$ and $x_2 \geq 0$ [29]. It suffices to show that the minimum of $f(x_1, x_2) = -(x_1^2 + x_2^2)e^{-x_1-x_2}$ is $-4e^{-2}$. According to Proposition 4.2.1 with $h_1(x_1, x_2) = -x_1$ and $h_2(x_1, x_2) = -x_2$, a minimum point necessarily satisfies

$$\begin{aligned} -\frac{\partial}{\partial x_1} f(x_1, x_2) &= (2x_1 - x_1^2 - x_2^2)e^{-x_1-x_2} \\ &= -\mu_1 \\ -\frac{\partial}{\partial x_2} f(x_1, x_2) &= (2x_2 - x_1^2 - x_2^2)e^{-x_1-x_2} \\ &= -\mu_2, \end{aligned}$$

where the multipliers μ_1 and μ_2 are nonnegative and satisfy $x_1\mu_1 = 0$ and $\mu_2 x_2 = 0$. In this problem, the Mangasarian-Fromovitz constraint qualification is trivial to check using the vector $v = \mathbf{1}$. If neither x_1 nor x_2 vanish, then

$$2x_1 - x_1^2 - x_2^2 = 2x_2 - x_1^2 - x_2^2 = 0.$$

This forces $x_1 = x_2$ and $2x_1 - 2x_1^2 = 0$. It follows that $x_1 = x_2 = 1$, where $f(1, 1) = -2e^{-2}$. We can immediately eliminate the origin $\mathbf{0}$ from contention because $f(0, 0) = 0$. If $x_1 = 0$ and $x_2 > 0$, then $\mu_2 = 0$ and $2x_2 - x_2^2 = 0$. This implies that $x_2 = 2$ and $(0, 2)$ is a candidate minimum point. By symmetry, $(2, 0)$ is also a candidate minimum point. At these two boundary points, $f(2, 0) = f(0, 2) = -4e^{-2}$, and this verifies the claimed minimum value. ∎

Example 4.2.3 *Application to Linear Programming*

The gradient of the Lagrangian

$$\mathcal{L}(x,\lambda,\mu) = z^*x + \sum_{i=1}^{p} \lambda_i\Big(\sum_{j=1}^{n} v_{ij}x_j - d_i\Big) - \sum_{j=1}^{q} \mu_j x_j$$

vanishes at the minimum of the linear function $f(x) = z^*x$ subject to the constraints $Vx = d$ and $x \geq \mathbf{0}$. Differentiating $\mathcal{L}(x,\lambda,\mu)$ with respect to x_j and setting the result equal to 0 gives $z_j + \sum_{i=1}^{p} \lambda_i v_{ij} - \mu_j = 0$. In vector notation this is just

$$z + V^*\lambda - \mu = \mathbf{0}$$

subject to the restrictions $\mu \geq \mathbf{0}$ and $\mu^* x = 0$. We will revisit linear programming in Chapter 11, where we discuss an interior point algorithm and define dual linear programs. ■

Example 4.2.4 *A Counterexample to the Multiplier Rule*

The Lagrange multiplier condition is necessary but not sufficient for a point to furnish a minimum. For example, consider the function $f(x) = x_1^3 - x_2$ subject to the constraint $h(x) = x_2 \leq 0$. The Lagrange multiplier condition

$$\nabla f(\mathbf{0}) = \begin{pmatrix} 0 \\ -1 \end{pmatrix} = -\nabla h(\mathbf{0})$$

holds, but the origin $\mathbf{0}$ fails to minimize $f(x)$. Indeed, the one-dimensional slice $x_1 \mapsto f(x_1, 0)$ has a saddle point at $x_1 = 0$. This function has no minimum subject to the inequality constraint. ■

Example 4.2.5 *Quadratic Programming with Equality Constraints*

To minimize the quadratic function $f(x) = \frac{1}{2}x^*Ax + b^*x + c$ subject to the linear equality constraints $Vx = d$, we introduce the Lagrangian

$$\begin{aligned}\mathcal{L}(x,\lambda) &= \frac{1}{2}x^*Ax + b^*x + \sum_{i=1}^{p} \lambda_i[v_{(i)}^* x - d_i] \\ &= \frac{1}{2}x^*Ax + b^*x + \lambda^*(Vx - d).\end{aligned}$$

A stationary point of $\mathcal{L}(x,\lambda)$ is determined by the equations

$$\begin{aligned}Ax + b + V^*\lambda &= \mathbf{0} \\ Vx &= d,\end{aligned}$$

whose formal solution amounts to

$$\begin{pmatrix} x \\ \lambda \end{pmatrix} = \begin{pmatrix} A & V^* \\ V & \mathbf{0} \end{pmatrix}^{-1} \begin{pmatrix} -b \\ d \end{pmatrix}.$$

The next proposition shows that the indicated matrix inverse exists when A is positive definite. ■

Proposition 4.2.2 *Let A be an $n \times n$ positive definite matrix and V be a $p \times n$ matrix. Then the matrix*

$$M = \begin{pmatrix} A & V^* \\ V & \mathbf{0} \end{pmatrix}$$

is invertible if and only if V has linearly independent rows $v_{(1)}^*, \ldots, v_{(p)}^*$. When this condition holds, M has inverse

$$M^{-1} = \begin{pmatrix} A^{-1} - A^{-1}V^*(VA^{-1}V^*)^{-1}VA^{-1} & A^{-1}V^*(VA^{-1}V^*)^{-1} \\ (VA^{-1}V^*)^{-1}VA^{-1} & -(VA^{-1}V^*)^{-1} \end{pmatrix}.$$

Proof: We first show that the symmetric matrix M is invertible with the specified inverse if and only if $(VA^{-1}V^*)^{-1}$ exists. If M^{-1} exists, it is necessarily symmetric. Suppose M^{-1} is given by $\begin{pmatrix} B & C^* \\ C & D \end{pmatrix}$. Then the identity

$$\begin{pmatrix} A & V^* \\ V & \mathbf{0} \end{pmatrix}\begin{pmatrix} B & C^* \\ C & D \end{pmatrix} = \begin{pmatrix} I_n & \mathbf{0} \\ \mathbf{0} & I_p \end{pmatrix}$$

implies that $VC^* = I_p$ and $AC^* + V^*D = \mathbf{0}$. Multiplying the last equality by VA^{-1} gives $I_p = -VA^{-1}V^*D$. Thus, $(VA^{-1}V^*)^{-1}$ exists. Conversely, if $(VA^{-1}V^*)^{-1}$ exists, then one can check by direct multiplication that M has the claimed inverse.

If $(VA^{-1}V^*)^{-1}$ exists, then V must have full row rank p. Conversely, if V has full row rank p, take any nontrivial $u \in \mathsf{R}^p$. Then the fact

$$u^*V = u_1v_{(1)}^* + \cdots u_pv_{(p)}^* \neq \mathbf{0}^*$$

and the positive definiteness of A imply $u^*VA^{-1}V^*u > 0$. Thus, $VA^{-1}V^*$ is positive definite and invertible. ■

4.3 Constraint Qualification

Fortunately, the Mangasarian-Fromovitz constraint qualification is a consequence of a simpler condition suggested by Kuhn and Tucker. Suppose the differentials $dg_i(x)$ and $dh_j(x)$ of the equality constraints and the active inequality constraints are linearly independent at x. If there are r active inequality constraints, then we can find an $n \times n$ invertible matrix A whose first p rows are the vectors $dg_i(x)$ and whose next r rows are the vectors $dh_1(x), \ldots, dh_r(x)$ corresponding to the active inequality constraints. For example, we can choose the last $n - p - r$ rows of A to be a basis for the orthogonal complement of the subspace spanned by the first $p + r$ rows of A. Given that A is invertible, there certainly exists a vector $v \in \mathsf{R}^n$ with

$$Av = -\begin{pmatrix} \mathbf{0} \\ \mathbf{1} \\ \mathbf{0} \end{pmatrix},$$

where the column vectors on the right side of this equality have p, r, and $n - p - r$ rows, respectively. Clearly, v satisfies the constraint qualification.

Even if the Kuhn-Tucker condition fails, there is still a chance that the constraint qualification holds. Let G be the $p \times n$ matrix with rows $dg_i(x)$. By assumption G has full rank. Taking the orthogonal complement of the subspace spanned by the p rows of G, we can construct an $n \times (n - p)$ matrix K of full rank whose columns are orthogonal to the rows of G. This fact can be expressed as $GK = \mathbf{0}$ and implies that the image of the linear transformation K from R^{n-p} to R^n is the kernel or null space of G. In other words, any v satisfying $Gv = \mathbf{0}$ can be expressed as $v = Ku$ and vice versa. If we let $z^*_{(j)} = dh_j(x)K$ for each of the active inequality constraints, then the Mangasarian-Fromovitz constraint qualification is equivalent to the existence of a vector $u \in \mathsf{R}^{n-p}$ satisfying $z^*_{(j)}u < 0$ for all $1 \leq j \leq r$.

The next proposition paves the way for proving a necessary and sufficient condition for this equality-free form of the constraint qualification. The result described by the proposition is of independent interest; its proof illustrates the fact that adding small penalties as opposed to large penalties is sometimes helpful [37].

Proposition 4.3.1 (Ekeland) *Suppose the real-valued function $f(x)$ is defined and differentiable on R^n. If $f(x)$ is bounded below, then there are points where $\|\nabla f(x)\|$ is arbitrarily close to 0.*

Proof: Take any small $\epsilon > 0$ and define the continuous function

$$f_\epsilon(x) = f(x) + \epsilon\|x\|.$$

In view of the boundedness condition, for any y the set $\{x : f_\epsilon(x) \leq f_\epsilon(y)\}$ is compact. Hence, Proposition 2.5.4 implies that $f_\epsilon(x)$ has a global minimum z depending on ϵ. If $v = \nabla f(z)$ satisfies $\|v\| > \epsilon$, then the limit relation

$$\begin{aligned}
\lim_{t \downarrow 0} \frac{f(z - tv) - f(z)}{t} &= -df(z)v \\
&= -\|v\|^2 \\
&< -\epsilon\|v\|,
\end{aligned}$$

the choice of z, and the triangle inequality together imply

$$\begin{aligned}
-t\epsilon\|v\| &> f(z - tv) - f(z) \\
&= f_\epsilon(z - tv) - f_\epsilon(z) + \epsilon\|z\| - \epsilon\|z - tv\| \\
&\geq \epsilon\|z\| - \epsilon\|z - tv\| \\
&\geq -\epsilon t\|v\|
\end{aligned}$$

for sufficiently small $t > 0$. This contradiction implies that $\|v\| \leq \epsilon$. ■

We are now in position to characterize the Mangasarian-Fromovitz constraint qualification in the absence of equality constraints. In essence, the constraint qualification holds if and only if the convex set generated by the active inequality constraints does not contain the origin. The geometric nature of this result will be clearer after we consider convex sets in the next chapter.

Proposition 4.3.2 (Gordon) *Given r vectors $z_{(1)}, \ldots, z_{(r)}$ in R^n, define the function $f(x) = \ln\left[\sum_{j=1}^r \exp(z_{(j)}^* x)\right]$. Then the following three conditions are logically equivalent:*

(a) *The function $f(x)$ is bounded below on R^n.*

(b) *There are nonnegative constants $\mu_1, \ldots, \mu_r$ such that*

$$\sum_{i=1}^r \mu_i z_{(i)} = \mathbf{0}, \qquad \sum_{i=1}^r \mu_i = 1.$$

(c) *There is no vector u such that $z_{(j)}^* u < 0$ for all j.*

Proof: It is trivial to check that (b) implies (c) and (c) implies (a). To demonstrate that (a) implies (b), first observe that

$$\nabla f(x) = \frac{1}{\sum_{i=1}^r e^{z_{(i)}^* x}} \sum_{j=1}^r e^{z_{(j)}^* x} z_{(j)}.$$

According to Proposition 4.3.1, it is possible to choose for each k a point $u_{(k)}$ at which

$$\|\nabla f(u_{(k)})\| = \left\|\sum_{j=1}^r \mu_{kj} z_{(j)}\right\| \leq \frac{1}{k},$$

where

$$\mu_{kj} = \frac{e^{z_{(j)}^* u_{(k)}}}{\sum_{i=1}^r e^{z_{(i)}^* u_{(k)}}}.$$

Because the μ_{kj} form a vector $\mu_{(k)}$ with nonnegative components summing to 1, it is possible to find a subsequence of the sequence $\mu_{(k)}$ that converges to a vector μ having the same properties. This vector satisfies the requirements of condition (b). ■

4.4 Taylor-Made Second Differentials

Suppose $f(y)$ is a real-valued function on the open domain $U \subset \mathsf{R}^m$. We have defined $f(y)$ to be differentiable at the point $x \in U$ provided there exists a slope function $s(y, x)$ such that

$$f(y) - f(x) = s(y, x)(y - x)$$

for all y near x and such that $\lim_{y \to x} s(y, x) = df(x) = \nabla f(x)^*$ exists. It now makes sense to postulate that the slope function $s(y, x)$ itself is differentiable at x. Because $s(y, x)$ is a row vector rather than a column vector, we require the existence of a second slope function $s^2(y, x)$ satisfying

$$s(y, x)^* - \nabla f(x) = \frac{1}{2} s^2(y, x)^* (y - x) \tag{4.7}$$

in a neighborhood of x. Of course, we retain the vital continuity assumption $\lim_{y \to x} s^2(y, x) = s^2(x, x)$. The primary virtue of this rather odd definition is that it leads to the exact Taylor expansion

$$f(y) = f(x) + df(x)(y - x) + \frac{1}{2}(y - x)^* s^2(y, x)(y - x) \tag{4.8}$$

for all y near x.

The definition of the second slope $s^2(y, x)$ also ensures that $s^2(x, x)$ is unique. Indeed, if $r(y, x)$ is another slope function with second slope $r^2(y, x)$, then it is obvious that

$$\begin{aligned} v^* r^2(x, x) v &= \lim_{t \to 0} \frac{2[f(x + tv) - f(x) - t df(x) v]}{t^2} \\ &= v^* s^2(x, x) v \end{aligned}$$

for all vectors v. In view of the arguments given in Example 2.5.7, the equality of the two quadratic forms $v^* r^2(x, x) v$ and $v^* s^2(x, x) v$ for all v implies the equality of the corresponding matrices $r^2(x, x)$ and $s^2(x, x)$, provided these are symmetric.

It is interesting to compare the expansion (4.8) to the more familiar expansion

$$\begin{aligned} f(y) &= f(x) + df(x)(y - x) \\ &\quad + (y - x)^* \int_0^1 d^2 f[x + t(y - x)](1 - t)\, dt\, (y - x) \end{aligned} \tag{4.9}$$

when $f(y)$ has continuous second-order partial derivatives in a neighborhood of x. As in the proof of Proposition 3.4.1, one can validate this expansion by integration by parts starting with $s(y, x) = \int_0^1 df[x + t(y - x)]dt$. Equation (4.9) correctly suggests that

$$s^2(y, x) = \int_0^1 2d^2 f[x + t(y - x)](1 - t)\, dt \tag{4.10}$$

is a valid second slope whose limit $d^2 f(x)$ is a symmetric matrix. In proceeding axiomatically, we prefer not to impose such strong assumptions on $f(x)$.

Our current second-slope definition does not readily lead to the existence of second partial derivatives. It would be helpful to adopt a definition that forces their existence. To achieve this end, it suffices to impose the symmetry conditions $s(y,x) = s(x,y)$ and

$$s(x,y)^* - \nabla f(y) = \frac{1}{2}s^2(x,y)^*(x-y) \tag{4.11}$$

and the limit condition $\lim_{y\to x} s^2(x,y) = s^2(x,x)$. Subtracting equation (4.11) from equation (4.7) and applying the symmetry conditions now yields

$$\nabla f(y) - \nabla f(x) = \frac{1}{2}\left[s^2(x,y) + s^2(y,x)\right]^* (y-x). \tag{4.12}$$

Thus, $\nabla f(y)$ is differentiable at x with symmetric slope function

$$s_{\nabla f}(y,x) = \frac{1}{2}\left[s^2(x,y) + s^2(y,x)\right]^*.$$

Provided $s^2(x,x)$ is a symmetric matrix, we are justified in regarding it as the second differential $d^2 f(x)$ of $f(x)$.

In view of these motivating remarks, we now define $f(y)$ to be twice differentiable at x provided (a) it is differentiable for all y in a neighborhood of x with slope function $s(y,x)$ symmetric in its two arguments, (b) it possesses second slope functions $s^2(y,x)$ and $s^2(x,y)$ satisfying

$$s(y,x)^* - \nabla f(x) = \frac{1}{2}s^2(y,x)^*(y-x) \tag{4.13}$$

$$s(x,y)^* - \nabla f(y) = \frac{1}{2}s^2(x,y)^*(x-y), \tag{4.14}$$

and (c) $\lim_{y\to x} s^2(y,x) = \lim_{y\to x} s^2(x,y) = d^2 f(x)$ exists and is a symmetric matrix. Equation (4.12) and assumption (c) imply that

$$\lim_{y\to x} \nabla f(y) = \nabla f(x).$$

We will call $f(y)$ twice continuously differentiable on an open set U if and only if it is twice differentiable and possesses jointly continuous second slope functions $s^2(y,x)$ and $s^2(x,y)$ near every point $z \in U$. In view of the canonical choice (4.10), this is equivalent to $d^2 f(x)$ being continuous on U.

Example 4.4.1 *Second Differential of a Quadratic Function*

Consider the quadratic function $f(x) = \frac{1}{2}x^*Ax + b^*x + c$ defined by an $n \times n$ symmetric matrix A and an $n \times 1$ vector b. The brief calculation

$$\begin{aligned} f(y) - f(x) &= \frac{1}{2}y^*Ay + b^*y + c - \frac{1}{2}x^*Ax - b^*x - c \\ &= \frac{1}{2}(y+x)^*A(y-x) + b^*(y-x) \end{aligned}$$

identifies the slope function $s(y, x) = \frac{1}{2}(y + x)^* A + b^*$ near x. The second slope function emerges as A from the expansion

$$\begin{aligned} s(y, x) - df(x) &= \frac{1}{2}(y + x)^* A + b^* - \frac{1}{2}(x + x)^* A - b^* \\ &= \frac{1}{2}(y - x)^* A. \end{aligned}$$

This yields the second differential A that we have already encountered. ■

Example 4.4.2 *A Pathological Example*

On R^2 define the indicator function $1_Q(x)$ to be 1 when both coordinates x_1 and x_2 are rational numbers and to be 0 otherwise. The function

$$f(x) = 1_Q(x)(x_1^3 + x_2^3)$$

is differentiable at the origin $\mathbf{0}$ but nowhere else on R^2. Its slope function

$$s(y, \mathbf{0}) = 1_Q(y)(y_1^2, y_2^2)$$

at the origin satisfies equation (4.13) with second slope function

$$s^2(y, \mathbf{0}) = 1_Q(y) \begin{pmatrix} 2y_1 & 0 \\ 0 & 2y_2 \end{pmatrix}.$$

Furthermore, $s^2(y, \mathbf{0})$ tends to the $\mathbf{0}$ matrix as y tends to the origin. However, claiming that $f(x)$ is twice differentiable at the origin seems extreme. To eliminate such pathologies, we demand that $f(x)$ be differentiable in a neighborhood of a point before we allow it to be twice differentiable at the point. ■

The preceding arguments settle the matter of an appropriate definition of the second differential except that we must lift the restriction to real-valued functions. If $f(y)$ has differentiable components $f_1(y), \ldots, f_n(y)$ with slope functions $s_{(1)}(y, x), \ldots, s_{(n)}(y, x)$, then the natural slope function $s(y, x)$ for $f(y)$ has these slope functions as rows. Assuming each component slope function is differentiable, we write

$$\begin{aligned} s(y, x) - df(x) &= \begin{pmatrix} s_{(1)}(y, x) - df_1(x) \\ \vdots \\ s_{(n)}(y, x) - df_n(x) \end{pmatrix} \\ &= \frac{1}{2} \begin{pmatrix} (y - x)^* & \mathbf{0} & \cdots & \mathbf{0} \\ \mathbf{0} & (y - x)^* & \cdots & \mathbf{0} \\ \vdots & \vdots & \ddots & \vdots \\ \mathbf{0} & \mathbf{0} & \cdots & (y - x)^* \end{pmatrix} \begin{pmatrix} s_{(1)}^2(y, x) \\ \vdots \\ s_{(n)}^2(y, x) \end{pmatrix} \\ &= \frac{1}{2}[I_n \otimes (y - x)^*] s^2(y, x), \end{aligned} \qquad (4.15)$$

where I_n is the $n \times n$ identity matrix and $s^2(y, x)$ is the block matrix with blocks $s^2_{(i)}(y, x)$. Formula (4.15) and its obvious analog reversing x and y serve as substitutes for formulas (4.13) and (4.14) in defining second-order differentiability for a vector-valued function.

Example 4.4.3 *Second Differential of the Inverse Polar Transformation*

Continuing Example 3.8.1, it is straightforward to calculate that the inverse polar transformation

$$g\left[\begin{pmatrix} r \\ \theta \end{pmatrix}\right] = \begin{pmatrix} r\cos\theta \\ r\sin\theta \end{pmatrix}$$

has second differential

$$d^2 g\left[\begin{pmatrix} r \\ \theta \end{pmatrix}\right] = \begin{pmatrix} 0 & -\sin\theta \\ -\sin\theta & -r\cos\theta \\ 0 & \cos\theta \\ \cos\theta & -r\sin\theta \end{pmatrix}.$$

The next proposition emphasizes the importance of second differentials in optimization.

Proposition 4.4.1 *Consider a real-valued function $f(y)$ with domain an open set $U \subset \mathsf{R}^n$. If $f(y)$ has a local minimum at x and is twice differentiable there, then the second differential $d^2 f(x)$ is positive semidefinite. Conversely, if x is a stationary point of $f(y)$ and $d^2 f(x)$ is positive definite, then x is a local minimum of $f(y)$. Similar statements hold for local maxima if we replace the modifiers positive semidefinite and positive definite by the modifiers negative semidefinite and negative definite. Finally, if x is a stationary point of $f(y)$ and $d^2 f(x)$ possesses both positive and negative eigenvalues, then x is neither a local minimum nor a local maximum of $f(y)$.*

Proof: Suppose x provides a local minimum of $f(y)$. For any unit vector v and $t > 0$ sufficiently small, the point $y = x + tv$ belongs to U and satisfies $f(y) \geq f(x)$. If we divide the expansion

$$\begin{aligned} 0 &\leq f(y) - f(x) \\ &= \frac{1}{2}(y - x)^* s^2(y, x)(y - x) \end{aligned}$$

by $t^2 = \|y - x\|^2$ and send t to 0, then it follows that

$$0 \leq \frac{1}{2} v^* d^2 f(x) v.$$

Because v is an arbitrary unit vector, the quadratic form $d^2 f(x)$ must be positive semidefinite.

On the other hand, if x is a stationary point of $f(x)$ and $d^2f(x)$ is positive definite, suppose x fails to be a local minimum. Then there exists a sequence of points $y_{(m)}$ tending to x and satisfying

$$0 \;>\; f(y_{(m)}) - f(x) \;=\; \frac{1}{2}(y_{(m)} - x)^* s^2(y_{(m)}, x)(y_{(m)} - x). \qquad (4.16)$$

Passing to a subsequence if necessary, we may assume that the unit vectors $v_{(m)} = (y_{(m)} - x)/\|y_{(m)} - x\|$ converge to a unit vector v. Dividing inequality (4.16) by $\|y_{(m)} - x\|^2$ and sending m to ∞ consequently yields $0 \geq v^* d^2f(x)v$, contrary to the hypothesis that $d^2f(x)$ is positive definite. This contradiction shows that x represents a local minimum.

To prove the final claim of the proposition, let μ be a nonzero eigenvalue with corresponding eigenvector v. Then the difference

$$\begin{aligned} f(x+tv) - f(x) &= \frac{t^2}{2}\{v^*d^2f(x)v + v^*[s^2(x+tv,x) - d^2f(x)]v\} \\ &= \frac{t^2}{2}\{\mu\|v\|^2 + v^*[s^2(x+tv,x) - d^2f(x)]v\} \end{aligned}$$

has the same sign as μ for t small. ■

Example 4.4.4 *Distinguishing Extrema from Saddle Points*

Consider the function

$$f(x) \;=\; \frac{1}{4}x_1^4 + \frac{1}{2}x_2^2 - x_1x_2 + x_1 - x_2$$

on R^2. It is obvious that

$$\nabla f(x) \;=\; \begin{pmatrix} x_1^3 - x_2 + 1 \\ x_2 - x_1 - 1 \end{pmatrix}, \quad d^2f(x) \;=\; \begin{pmatrix} 3x_1^2 & -1 \\ -1 & 1 \end{pmatrix}.$$

Adding the two rows of the stationarity equation $\nabla f(x) = \mathbf{0}$ gives the equation $x_1^3 - x_1 = 0$ with solutions $0, \pm 1$. Solving for x_2 in each case yields the stationary points $(0,1)$, $(-1,0)$, and $(1,2)$. The last two points are local minima because $df(x)$ is positive definite. The first point is a saddle point because

$$d^2f(0,1) \;=\; \begin{pmatrix} 0 & -1 \\ -1 & 1 \end{pmatrix}$$

has characteristic polynomial $\lambda^2 - \lambda - 1$ and eigenvalues $\frac{1}{2}(1 \pm \sqrt{5})$. One of these eigenvalues is positive, and one is negative. ■

The rules for calculating second differentials are naturally more complicated than those for calculating first differentials.

Proposition 4.4.2 *If the two functions $f(x)$ and $g(x)$ map the open set $S \subset \mathsf{R}^m$ twice differentiably into R^n, then the following functional combinations are twice differentiable and have the indicated second differentials:*

(a) *For all constants* α *and* β,

$$d^2[\alpha f(x) + \beta g(x)] = \alpha d^2 f(x) + \beta d^2 g(x).$$

(b) *The inner product* $f(x)^* g(x)$ *satisfies*

$$\begin{aligned} d^2[f(x)^* g(x)] &= \sum_{i=1}^n \left[f_i(x) d^2 g_i(x) + g_i(x) d^2 f_i(x) \right] \\ &\quad + \sum_{i=1}^n \left[df_i(x)^* dg_i(x) + dg_i(x)^* df_i(x) \right]. \end{aligned}$$

(c) *When* $n = 1$ *and* $f(x) \neq 0$,

$$d^2[f(x)^{-1}] = -f(x)^{-2} d^2 f(x) + 2f(x)^{-3} df(x)^* df(x).$$

Proof: Let the slope functions for $f(x)$ and $g(x)$ at x be $s_f(y,x)$ and $s_g(y,x)$. Rule (a) follows directly from the linearity implicit in formula (4.15). For part (b) it suffices to consider the scalar case in view of part (a). Using the symmetric slope function and differential identified in equation (3.11), we have

$$\begin{aligned} & \frac{1}{2}[f(y) + f(x)] s_g(y,x) + \frac{1}{2}[g(y) + g(x)] s_f(y,x) \\ & - f(x) dg(x) - g(x) df(x) \\ = \; & \frac{1}{2}[f(y) + f(x)][s_g(y,x) - dg(x)] + \frac{1}{2}[g(y) + g(x)][s_f(y,x) - df(x)] \\ & + \frac{1}{2}[f(y) - f(x)] dg(x) + \frac{1}{2}[g(y) - g(x)] df(x) \\ = \; & \frac{1}{2}[f(y) + f(x)] \frac{1}{2}(y-x)^* s_g^2(y,x) + \frac{1}{2}[g(y) + g(x)] \frac{1}{2}(y-x)^* s_f^2(y,x) \\ & + \frac{1}{2}(y-x)^* s_f(y,x)^* dg(x) + \frac{1}{2}(y-x)^* s_g(y,x)^* df(x). \end{aligned}$$

It follows that the product $f(y)g(y)$ has second slope

$$\begin{aligned} s_{fg}^2(y,x) &= \frac{1}{2}[f(y) + f(x)] s_g^2(y,x) + \frac{1}{2}[g(y) + g(x)] s_f^2(y,x) \\ &\quad + s_f(y,x)^* dg(x) + s_g(y,x)^* df(x). \end{aligned}$$

A similar calculation gives

$$\begin{aligned} s_{fg}^2(x,y) &= \frac{1}{2}[f(y) + f(x)] s_g^2(x,y) + \frac{1}{2}[g(y) + g(x)] s_f^2(x,y) \\ &\quad + s_f(x,y)^* dg(y) + s_g(x,y)^* df(y). \end{aligned}$$

Both of these second slopes yield in the limit the claimed symmetric second differential.

To verify rule (c), we use the slope function and differential for $f(y)^{-1}$ identified in the proof of Proposition 3.6.1. Together these enable us to calculate

$$\begin{aligned}
& -\frac{1}{f(y)f(x)}s_f(y,x) + \frac{1}{f(x)^2}df(x) \\
= & -\frac{1}{f(y)f(x)}[s_f(y,x) - df(x)] + \frac{1}{f(y)f(x)^2}[f(y) - f(x)]df(x) \\
= & -\frac{1}{f(y)f(x)}\frac{1}{2}(y-x)^* s_f^2(y,x) + \frac{1}{f(y)f(x)^2}s_f(y,x)(y-x)df(x) \\
= & \frac{1}{2}(y-x)^* \frac{1}{f(y)f(x)}\left[-s_f^2(y,x) + \frac{2}{f(x)}s_f(y,x)^* df(x)\right].
\end{aligned}$$

Interchanging x and y gives the other second slope. Again, both of these produce in the limit the displayed symmetric second differential. ■

The chain rule also is more complex.

Proposition 4.4.3 *Suppose $f(x)$ maps the open set $S \subset \mathsf{R}^k$ twice differentiably into R^m and $g(y)$ maps the open set $T \subset \mathsf{R}^m$ twice differentiably into R^n. If the image $f(S)$ is contained in T, then the composition $g \circ f(x)$ is twice differentiable with second differential*

$$d^2 g \circ f = \begin{pmatrix} df(x)^* d^2 g_1 \circ f(x) df(x) + \sum_{i=1}^m d_i g_1 \circ f(x) d^2 f_i(x) \\ \vdots \\ df(x)^* d^2 g_n \circ f(x) df(x) + \sum_{i=1}^m d_i g_n \circ f(x) d^2 f_i(x) \end{pmatrix}.$$

Proof: It suffices to prove the result when $n = 1$ and $g(x)$ is scalar valued. Let the slope functions for $f(x)$ and $g(x)$ at x and $z = f(x)$ be $s_f(y,x)$ and $s_g(y,z)$. To exhibit a second slope, we calculate

$$\begin{aligned}
& s_g[f(y), f(x)]s_f(y,x) - dg \circ f(x) df(x) \\
= & [s_g[f(y), f(x)] - dg \circ f(x)]s_f(y,x) + dg \circ f(x)[s_f(y,x) - df(x)] \\
= & \frac{1}{2}[f(y) - f(x)]^* s_g^2[f(y), f(x)]s_f(y,x) \\
& + \sum_{i=1}^m d_i g \circ f(x)[s_{f_i}(y,x) - df_i(x)] \\
= & \frac{1}{2}(y-x)^* s_f(y,x)^* s_g^2[f(y), f(x)]s_f(y,x) \\
& + \sum_{i=1}^m d_i g \circ f(x) \frac{1}{2}(y-x)^* s_{f_i}^2(y,x) \\
= & \frac{1}{2}(y-x)^* \left[s_f(y,x)^* s_g^2[f(y), f(x)]s_f(y,x) + \sum_{i=1}^m d_i g \circ f(x) s_{f_i}^2(y,x)\right]
\end{aligned}$$

based on the slope function used in the proof of Proposition 3.6.2. Taking limits gives the desired second differential. ■

The last proposition of this section deals with second differentials and the inverse function theorem.

Proposition 4.4.4 *If the function $f(x)$ in the inverse function theorem as stated and proved in Proposition 3.8.1 is twice differentiable at x, then the inverse function $g(y)$ is twice differentiable at $f(x)$ and has second differential*

$$d^2g(x) = -[df \circ g(x)^{-1} \otimes I_n] \begin{pmatrix} [df \circ g(x)^{-1}]^* df_1^2 \circ g(x)\, df \circ g(x)^{-1} \\ \vdots \\ [df \circ g(x)^{-1}]^* df_n^2 \circ g(x)\, df \circ g(x)^{-1} \end{pmatrix}.$$

Proof: See Problem 19. ■

4.5 A Sufficient Condition for a Minimum

We now state and prove the promised sufficient condition for a point x to be a local minimum of the objective function $f(y)$. Even in the absence of constraints, inequality (4.17) below represents an improvement over the qualitative claims of Proposition 4.4.1.

Proposition 4.5.1 *Suppose the objective function $f(y)$ of the constrained optimization problem satisfies the multiplier rule (4.1) at the point x with $\lambda_0 = 1$. Let $f(y)$ and the various constraint functions be twice differentiable at x, and let $F(y)$ be the Lagrangian*

$$F(y) = f(y) + \sum_{i=1}^{p} \lambda_i g_i(y) + \sum_{j=1}^{q} \mu_j h_j(y).$$

If $v^* d^2F(x)v > 0$ for every vector $v \neq \mathbf{0}$ satisfying $dg_i(x)v = 0$ and $dh_j(x)v \leq 0$ for all active constraints, then x provides a local minimum of $f(y)$. Furthermore, there exists a constant $c > 0$ such that

$$f(y) \geq f(x) + c\|y - x\|^2 \tag{4.17}$$

for all feasible y in a neighborhood of x.

Proof: Because of the sign restrictions on the μ_j, we have $f(y) \geq F(y)$ for all feasible y in addition to $f(x) = F(x)$. If $F(y) \geq F(x)$, then $f(y) \geq f(x)$, and if $F(y) \geq F(x) + c\|y - x\|^2$, then $f(y) \geq f(x) + c\|y - x\|^2$. Therefore, it suffices to prove these inequalities for $F(y)$ rather than $f(y)$. The second inequality $F(y) \geq F(x) + c\|y - x\|^2$ is stronger than the first inequality $F(y) \geq F(x)$, so it also suffices to focus on the second inequality.

With this end in mind, let $s_F^2(y,x)$ be a second slope function for $F(y)$ at x. If the second inequality is false, then there exists a sequence of feasible points $y_{(m)}$ converging to x and a sequence of positive constants c_m converging to 0 such that

$$\begin{aligned} F(y_{(m)}) - F(x) &= \frac{1}{2}(y_{(m)} - x)^* s_F^2(y_{(m)}, x)(y_{(m)} - x) \\ &< c_m \|y_{(m)} - x\|^2. \end{aligned} \tag{4.18}$$

Here $dF(x)$ vanishes by virtue of the multiplier condition. As usual, we suppose that the sequence of unit vectors

$$v_{(m)} = \frac{1}{\|y_{(m)} - x\|}(y_{(m)} - x)$$

converges to a unit vector v by extracting a subsequence if necessary. Dividing inequality (4.18) by $\|y_{(m)} - x\|^2$ and taking limits then yields $v^* d^2F(x)v \le 0$. This contradicts our supposition about $d^2F(x)$ provided we can demonstrate that the tangent conditions $dg_i(x)v = 0$ and $dh_j(x)v \le 0$ hold for all active constraints. These follow by dividing the equations

$$0 = g_i(y_{(m)}) - g_i(x) = s_{g_i}(y_{(m)}, x)(y_{(m)} - x)$$

and

$$0 \ge h_j(y_{(m)}) - h_j(x) = s_{h_j}(y_{(m)}, x)(y_{(m)} - x)$$

by $\|y_{(m)} - x\|$ and taking limits. Recall here that $h_j(x) = 0$ at an active constraint. ∎

Example 4.5.1 *Minimum Eigenvalue of a Symmetric Matrix*

Example 1.4.3 demonstrated how each eigenvector-eigenvalue pair (x, α) of a symmetric matrix M provides a stationary point of a Lagrangian

$$F(x) = \frac{1}{2}x^* Mx - \frac{\alpha}{2}(\|x\|^2 - 1).$$

Suppose that the eigenvalues are arranged so that $\alpha_1 \le \cdots \le \alpha_n$ and $x_{(i)}$ is the eigenvector corresponding to α_i. We expect that $x_{(1)}$ furnishes the minimum of $\frac{1}{2}y^* My$ subject to $g_1(y) = \frac{1}{2} - \frac{1}{2}\|y\|^2 = 0$. To check that this is indeed the case, we note that $d^2F(y) = M - \alpha_1 I_n$. The condition $dg_1(x_{(1)})v = 0$ is equivalent to $x_{(1)}^* v = 0$. This can hold only if

$$v = \sum_{i=2}^{n} c_i x_{(i)}$$

owing to the orthogonality of the eigenvectors. For such a choice of v, the quadratic form

$$v^* d^2 F(x_{(1)}) v = \sum_{i=2}^{n} c_i^2 (\alpha_i - \alpha_1) > 0$$

so long as $\alpha_1 < \alpha_2$ and $v \neq \mathbf{0}$. Thus, the condition cited in the statement of Proposition 4.5.1 holds, and the point $x_{(1)}$ minimizes $y^* M y$ subject to the constraint $\|y\| = 1$. ■

Example 4.5.2 *Minimum of a Linear Reciprocal Function*

Consider the problem of minimizing the linear function $f(x) = \sum_{i=1}^n c_i x_i^{-1}$ subject to the linear inequality constraint $\sum_{i=1}^n a_i x_i \leq b$. Here all indicated variables and parameters are positive. Differentiating the Lagrangian

$$F(x) = \sum_{i=1}^{n} c_i x_i^{-1} + \mu \left(\sum_{i=1}^{n} a_i x_i - b \right)$$

gives the multiplier equations

$$-\frac{c_i}{x_i^2} + \mu a_i = 0.$$

It follows that $\mu > 0$, that the constraint is active, and that

$$\begin{aligned} x_i &= \sqrt{\frac{c_i}{\mu a_i}}, \qquad 1 \leq i \leq n \\ \mu &= \left(\frac{1}{b} \sum_{i=1}^{n} \sqrt{a_i c_i} \right)^2 . \end{aligned} \tag{4.19}$$

The second differential $d^2 F(x)$ is diagonal with ith diagonal entry $2c_i/x_i^3$. This matrix is certainly positive definite, and Proposition 4.5.1 confirms that the stationary point (4.19) provides the minimum of $f(x)$ subject to the constraint. ■

When there are only equality constraints, one can say more about the sufficient criterion described in Proposition 4.5.1. Following the discussion in Section 4.3, let G be the $p \times n$ matrix G with rows $dg_i(x)$ and K an $n \times (n-p)$ matrix of full rank satisfying $GK = \mathbf{0}$. On the kernel of G the matrix $A = d^2 F(x)$ is positive definite. Since every v in the kernel equals some image point Ku, we can establish the validity of the sufficient condition of Proposition 4.5.1 by checking whether the matrix $K^* A K$ of the quadratic form $u^* K^* A K u$ is positive definite. There are many practical methods of making this determination. For instance, the sweep operator from computational statistics performs such a check easily in the process of inverting $K^* A K$ [82].

If we want to work directly with the matrix G, there is another interesting criterion involving the relation between the positive semidefinite matrix $B = G^*G$ and the second differential $A = d^2F(x)$. One can rephrase the sufficient condition of Proposition 4.5.1 by saying that $v^*Av > 0$ whenever $v^*Bv = 0$ and $v \neq \mathbf{0}$. We claim that this condition is equivalent to the existence of some constant $\gamma > 0$ such that the matrix $A + \gamma B$ is positive definite [23]. Clearly, if such a constant exists, then the condition holds. Conversely, suppose the condition holds and that no such γ exists. Then there is a sequence of unit vectors $v_{(m)}$ and a sequence of scalars α_m tending to ∞ such that

$$v_{(m)}^* A v_{(m)} + \alpha_m v_{(m)}^* B v_{(m)} \quad \leq \quad 0. \tag{4.20}$$

By passing to a subsequence if needed, we may assume that the sequence $v_{(m)}$ converges to a unit vector v. On one hand, because B is positive semidefinite, inequality (4.20) compels the conclusions $v_{(m)}^* A v_{(m)} \leq 0$, which must carry over to the limit. On the other hand, dividing inequality (4.20) by α_m and taking limits imply $v^*Bv \leq 0$ and therefore $v^*Bv = 0$. Because the limit vector v violates the condition $v^*Av > 0$, the required $\gamma > 0$ exists.

4.6 Problems

1. Find a minimum of $f(x) = x_1^2 + x_2^2$ subject to the inequality constraints $h_1(x) = -2x_1 - x_2 + 10 \leq 0$ and $h_2(x) = -x_1 \leq 0$. Prove that it is the global minimum.

2. Minimize the function $f(x) = e^{-(x_1+x_2)}$ subject to the constraints $h_1(x) = e^{x_1} + e^{x_2} - 20 \leq 0$ and $h_2(x) = -x_1 \leq 0$ on R^2.

3. Find the minimum and maximum of the function $f(x) = x_1 + x_2$ over the subset of R^2 defined by the constraints $h_i(x) \leq 0$ for

$$\begin{aligned} h_1(x) &= -x_1 \\ h_2(x) &= -x_2 \\ h_3(x) &= 1 - x_1 x_2. \end{aligned}$$

4. Consider the problem of minimizing $f(x) = (x_1 + 1)^2 + x_2^2$ subject to the inequality constraint $h(x) = -x_1^3 + x_2^2 \leq 0$ on R^2. Solve the problem by sketching the feasible region and using a little geometry. Show that the multiplier rule of Proposition 4.2.1 with $\lambda_0 = 1$ fails and explain why.

5. Consider the inequality constraint functions

$$\begin{aligned} h_1(x) &= -x_1 \\ h_2(x) &= -x_2 \\ h_3(x) &= x_1^2 + 4x_2^2 - 4 \\ h_4(x) &= (x_1 - 2)^2 + x_2^2 - 5. \end{aligned}$$

Show that the Kuhn-Tucker constraint qualification fails but the Mangasarian-Fromovitz constraint qualification succeeds at the point $x = (0,1)^*$. For the inequality constraint functions

$$\begin{aligned} h_1(x) &= x_1^2 - x_2 \\ h_2(x) &= -3x_1^2 + x_2, \end{aligned}$$

show that both constraint qualifications fail at the point $x = (0,0)^*$ [46].

6. For a real 2×2 matrix

$$M = \begin{pmatrix} a & b \\ c & d \end{pmatrix},$$

define the Euclidean norm $\|M\|_E = \sqrt{a^2 + b^2 + c^2 + d^2}$. Let S denote the set of matrices M with $\det(M) = 0$. Find the minimum distance from S to the matrix

$$\begin{pmatrix} 1 & 0 \\ 0 & 2 \end{pmatrix},$$

and exhibit a matrix attaining this distance [29]. (Hints: Introduce a Lagrange multiplier λ in minimizing $\frac{1}{2}\|M\|_E^2$ subject to $M \in S$. From the multiplier conditions deduce that $\lambda = \pm 1$ if $b \neq 0$. Show that the assumption $\lambda = \pm 1$ leads to a contradiction. Thus, $b = 0$ and consequently $c = 0$. Express a and d as functions of λ and find the λ's for which $\det(M) = 0$.)

7. Let A be a positive definite matrix. For a given vector y, find the maximum of $f(x) = y^*x$ subject to $h(x) = x^*Ax - 1 \leq 0$. Use your result to prove the inequality $|y^*x|^2 \leq (x^*Ax)(y^*A^{-1}y)$.

8. A random variable takes the value x_i with probability p_i for i ranging from 1 to n. Maximize the entropy $-\sum_{i=1}^n p_i \ln p_i$ subject to a fixed mean $m = \sum_{i=1}^n x_i p_i$. Show that $p_i = \alpha e^{\lambda x_i}$ for constants α and λ. Argue that λ is determined by the equation

$$\sum_{i=1}^n x_i e^{\lambda x_i} = m \sum_{i=1}^n e^{\lambda x_i}.$$

9. Continuing the previous problem, suppose that each $x_j = j$. At the maximum, show that

$$p_i = \frac{p^{i-1}(1-p)}{1-p^n}$$

for some $p > 0$ and all $1 \leq i \leq n$. Argue that p exists and is unique for $n > 1$.

10. Consider the problem of minimizing the continuously differentiable function $f(x)$ subject to the constraint $x \geq \mathbf{0}$. At a local minimum y demonstrate that the partial derivative $d_i f(y) = 0$ when $y_i > 0$ and $d_i f(y) \geq 0$ when $y_i = 0$.

11. As a variation on Problem 10, consider minimizing the continuously differentiable function $f(x)$ subject to the constraints $\sum_{i=1}^n x_i = 1$ and $x \geq \mathbf{0}$. At a local minimum y demonstrate that there exists a number λ such that the partial derivative $d_i f(y) = \lambda$ when $y_i > 0$ and $d_i f(y) \geq \lambda$ when $y_i = 0$. This result is known as Gibbs' lemma.

12. Find the minimum value of $f(x) = \|x\|^2$ subject to the constraints $\sum_{i=1}^n x_i = 1$ and $x \geq \mathbf{0}$. Interpret the result geometrically.

13. For $p > 1$ define the norm $\|x\|_p$ on R^n satisfying $\|x\|_p^p = \sum_{i=1}^n |x_i|^p$. For a fixed vector z, maximize $f(x) = z^*x$ subject to $\|x\|_p^p \leq 1$. Deduce Hölder's inequality $|z^*x| \leq \|x\|_p \|z\|_q$ for q defined by the equation $p^{-1} + q^{-1} = 1$.

14. Suppose A is an $n \times n$ positive definite matrix. Find the minimum of $f(x) = \frac{1}{2}x^*Ax$ subject to the constraint $z^*x - c \leq 0$. It may help to consider the cases $c \geq 0$ and $c < 0$ separately.

15. Suppose that $v_{(1)}, \ldots, v_{(m)}$ are orthogonal eigenvectors of the $n \times n$ symmetric matrix M. Subject to the constraints

$$\begin{aligned} \frac{1}{2}\|x\|^2 &= 1 \\ v_{(i)}^* x &= 0, \quad 1 \leq i \leq m < n, \end{aligned}$$

show that a minimum of $\frac{1}{2}x^*Mx$ must coincide with an eigenvector of M. Under what circumstances is there a unique minimum of x^*Mx subject to the constraints?

16. Consider the set of $n \times n$ matrices $M = (m_{ij})$. Demonstrate that $\det M$ has maximum value $\prod_{i=1}^n d_i$ subject to the constraints

$$\sqrt{\sum_{j=1}^n m_{ij}^2} = d_i$$

for $1 \leq i \leq n$. This is Hadamard's inequality. (Hints: Use the Lagrange multiplier condition and the identities $\det M = \sum_{j=1}^n m_{ij} M_{ij}$ and $(M^{-1})_{ij} = M_{ji}/\det M$ to show that M can be written as the product DR of a diagonal matrix D with diagonal entries d_i and an orthogonal matrix R with $\det R = 1$.)

17. Find an explicit second slope for the function $f(x) = x^m$ on R. Here m is a positive integer.

18. Let $f(y)$ be a real-valued function of the real variable y. Suppose that $f''(y)$ exists at a point x. Prove that

$$f''(x) = \lim_{u \to 0} \frac{f(x+u) - 2f(x) + f(x-u)}{u^2}.$$

Use Problem 1 of Chapter 3 to devise an example where this limit quotient exists but $f''(x)$ does not exist.

19. Prove Proposition 4.4.4. (Hints: Verify and use the technical identities

$$A[I_n \otimes v^*] \begin{pmatrix} B_{(1)} \\ \vdots \\ B_{(n)} \end{pmatrix} = (I_n \otimes v^*)(A \otimes I_n) \begin{pmatrix} B_{(1)} \\ \vdots \\ B_{(n)} \end{pmatrix}$$

and

$$(A \otimes I_n) \begin{pmatrix} B_{(1)} \\ \vdots \\ B_{(n)} \end{pmatrix} = \begin{pmatrix} \sum_{j=1}^n a_{1j} B_{(j)} \\ \vdots \\ \sum_{j=1}^n a_{nj} B_{(j)} \end{pmatrix}$$

connecting the $n \times n$ matrices $A, B_{(1)}, \ldots, B_{(n)}$ and the $n \times 1$ vector v. The second identity helps in establishing the symmetry of the second differential.)

5
Convexity

5.1 Introduction

Convexity is one of the key concepts of mathematical analysis and has interesting consequences for optimization theory, statistical estimation, inequalities, and applied probability. Despite this fact, students seldom see convexity presented in a coherent fashion. It always seems to take a backseat to more pressing topics. The current chapter is intended as a partial remedy to this pedagogical gap.

We start with convex sets and proceed to convex functions. These intertwined concepts define and illuminate all sorts of inequalities. It is helpful to have a variety of tests to recognize convex functions. We present such tests and discuss the important class of log-convex functions. A strictly convex function has at most one minimum point. This property tremendously simplifies optimization. For a few functions, we are fortunate enough to be able to find their optima explicitly. For other functions, we must iterate.

The concluding section of this chapter rigorously treats several inequalities from the perspective of probability theory. Our inclusion of Bernstein's proof of the Weierstrass approximation theorem provides a surprising application of Chebyshev's inequality and illustrates the role of probability theory in solving problems outside its usual sphere of influence. The less familiar inequalities of Jensen, Schlömilch, and Hölder find numerous applications in optimization theory and functional analysis.

5.2 Convex Sets

A set $S \subset \mathsf{R}^m$ is said to be convex if the line segment between any two points x and y of S lies entirely within S. Formally, this means that whenever $x, y \in S$ and $\alpha \in [0,1]$, the point $z = \alpha x + (1-\alpha)y \in S$. In general, any convex combination $\sum_{i=1}^n \alpha_i x_i$ of points $x_1, \ldots, x_n$ in S must also reside in S. Here, the coefficients α_i must be nonnegative and sum to 1. It is easy to concoct examples of convex sets. For example, every interval on the real line is convex; every ball in R^n, either open or closed, is convex; and every multidimensional rectangle, either open or closed, is convex.

These examples suggest that the closure of a convex set is convex. To prove this fact, consider two points x and y in the closure of a convex set S. There exist sequences $u_{(k)}$ and $v_{(k)}$ from S converging to x and y, respectively. The convex combination $w_{(k)} = \alpha u_{(k)} + (1-\alpha)v_{(k)}$ is in S as well and converges to $z = \alpha x + (1-\alpha)y$ in the closure of S. Convex sets have many other important properties. For instance, because a convex set is arcwise connected, it is connected. A third property is the uniqueness of best approximations.

Proposition 5.2.1 *For a convex set S of R^n, there is at most one point $y \in S$ attaining the minimum distance $d(x, S)$. If S is closed, there is exactly one point.*

Proof: These claims are obvious if x is in S. Suppose that x is not in S and that y and z in S both attain the minimum. Then $(y+z)/2 \in S$ and

$$\begin{aligned} d(x,S) &\le \|x - \frac{1}{2}(y+z)\| \\ &\le \frac{1}{2}\|x-y\| + \frac{1}{2}\|x-z\| \\ &= d(x,S). \end{aligned}$$

Hence, equality must hold in the displayed triangle inequality. This is possible if and only if $x - y = c(x - z)$ for some positive number c. In view of the fact that $d(x,S) = \|x-y\| = \|x-z\|$, the value of c is 1 and y and z coincide. The second assertion follows from Example 2.5.5 of Chapter 2. ■

Yet a fourth important property relates to separation by hyperplanes.

Proposition 5.2.2 *Consider a closed convex set S of R^n and a point x outside S. There exists a vector v and real number c such that*

$$v^*x \;>\; c \;\ge\; v^*z$$

for all $z \in S$. As a consequence, S equals the intersection of all closed halfspaces containing it. If x is a boundary point of S, then there exists a unit vector v such that $v^*x \ge v^*z$ for all $z \in S$.

Proof: Let y be the closest point to x in S. Suppose that we can prove the inequality

$$(x-y)^*(z-y) \quad \leq \quad 0 \tag{5.1}$$

for all $z \in S$. If we take $v = x - y$, then $c = v^*y \geq v^*z$ for all $z \in S$. Furthermore, $v^*x > v^*y = c$ because $v^*v = \|v\|^2 > 0$.

To prove inequality (5.1), suppose it fails. Then $(x-y)^*(z-y) > 0$ for some $z \in S$. For each $0 < \alpha < 1$, the point $\alpha z + (1-\alpha)y$ is in S, and

$$\begin{aligned} \|x - \alpha z - (1-\alpha)y\|^2 &= \|x - y - \alpha(z-y)\|^2 \\ &= \|x-y\|^2 - \alpha\left[2(x-y)^*(z-y) - \alpha\|z-y\|^2\right]. \end{aligned}$$

For α sufficiently small, the term above in square brackets is positive, so $\alpha z + (1-\alpha)y$ improves on the choice of y. This contradiction demonstrates inequality (5.1).

If x is a boundary point of S, then there exists a sequence of points $x_{(i)}$ outside S that converge to x. Let $v_{(i)}$ be the vector defining the hyperplane separating $x_{(i)}$ from S. Without loss of generality, we can assume that the sequence $v_{(i)}$ consists of unit vectors and that some subsequence $v_{(i_j)}$ converges to a unit vector v. Taking limits in the strict inequality $v_{(i_j)}^* x_{(i_j)} > v_{(i_j)}^* z$ for $z \in S$ then yields the desired result $v^*x \geq v^*z$. ■

5.3 Convex Functions

Convex functions are defined on convex sets. A real-valued function $f(x)$ defined on a convex set S is convex provided

$$f[\alpha x + (1-\alpha)y] \quad \leq \quad \alpha f(x) + (1-\alpha)f(y) \tag{5.2}$$

for all $x, y \in S$ and $\alpha \in [0,1]$. Figure 5.1 depicts how in one dimension definition (5.2) requires the chord connecting two points on the curve $f(x)$ to lie above the curve. If strict inequality holds in (5.2) for every $x \neq y$ and $\alpha \in (0,1)$, then $f(x)$ is said to be strictly convex. One can prove by induction that inequality (5.2) extends to

$$f\Big(\sum_{i=1}^n \alpha_i x_i\Big) \quad \leq \quad \sum_{i=1}^n \alpha_i f(x_i)$$

for any convex combination of points from S. A concave function satisfies the reverse of inequality (5.2).

Example 5.3.1 *Affine Functions Are Convex*

For an affine function $f(x) = a^*x + b$, equality holds in inequality (5.2). ■

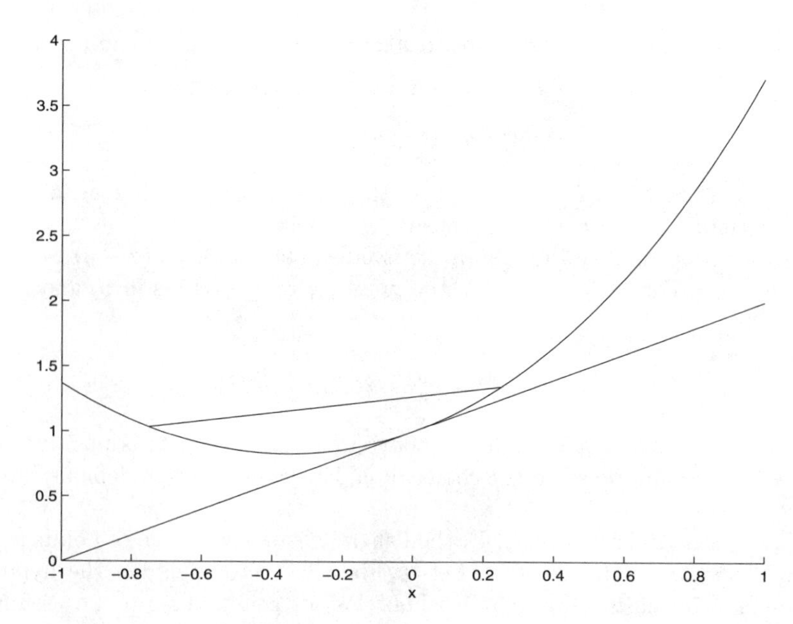

FIGURE 5.1. Plot of the Convex Function $e^x + x^2$

Example 5.3.2 *Norms Are Convex*

The Euclidean norm $f(x) = \|x\| = \sqrt{\sum_{i=1}^m x_i^2}$ satisfies the triangle inequality and the homogeneity condition $\|cx\| = |c|\,\|x\|$. Thus,

$$\|\alpha x + (1-\alpha)y\| \;\leq\; \|\alpha x\| + \|(1-\alpha)y\| \;=\; \alpha\|x\| + (1-\alpha)\|y\|$$

for any $\alpha \in [0,1]$. The same argument works for any norm. The choice $y = 2x$ gives equality in inequality (5.2) and shows that no norm is strictly convex. ■

Example 5.3.3 *The Distance to a Convex Set Is Convex*

The distance function $d(x,S)$ from points $x \in \mathsf{R}^n$ to a convex set S is convex. For any convex combination $\alpha x + (1-\alpha)y$, take sequences $u_{(k)}$ and $v_{(k)}$ from S such that

$$\begin{aligned} d(x,S) &= \lim_{k\to\infty} \|x - u_{(k)}\| \\ d(y,S) &= \lim_{k\to\infty} \|y - v_{(k)}\|. \end{aligned}$$

The points $\alpha u_{(k)} + (1-\alpha)v_{(k)}$ lie in S, and taking limits in the inequality

$$\begin{aligned} d[\alpha x + (1-\alpha)y, S] &\leq \|\alpha x + (1-\alpha)y - \alpha u_{(k)} - (1-\alpha)v_{(k)}\| \\ &\leq \alpha\|x - u_{(k)}\| + (1-\alpha)\|y - v_{(k)}\| \end{aligned}$$

yields $d[\alpha x + (1-\alpha)y, S] \leq \alpha d(x,S) + (1-\alpha)d(y,S)$. ■

Example 5.3.4 *Convex Functions Generate Convex Sets*

Consider a convex function $f(x)$ defined on R^n. Examination of definition (5.2) shows that the sets $\{x : f(x) \leq c\}$ and $\{x : f(x) < c\}$ are convex for any constant c. They may be empty. Conversely, a closed convex set S can be represented as $\{x : f(x) \leq 0\}$ using the continuous convex function $f(x) = d(x, S)$. ■

Example 5.3.5 *A Convex Function Has a Convex Epigraph*

The epigraph of a real-valued function $f(x)$ is defined as the set of points $(y^*, r)^*$ with $f(y) \leq r$. Roughly speaking, the epigraph is the region lying above the graph of $f(x)$. Consider two points $(y^*, r)^*$ and $(z^*, s)^*$ in the epigraph of $f(x)$. If $f(x)$ is convex, then

$$\begin{aligned} f[\alpha y + (1-\alpha)z] &\leq \alpha f(y) + (1-\alpha)f(z) \\ &\leq \alpha r + (1-\alpha)s, \end{aligned}$$

and the convex combination $\alpha(y^*, r)^* + (1-\alpha)(z^*, s)^*$ occurs in the epigraph of $f(x)$. ■

The next fact is much deeper.

Proposition 5.3.1 *A convex function $f(x)$ is continuous on the interior of its domain.*

Proof: Let y be an interior point and $C(y, r)$ be a closed ball of radius r around y contained within the domain of $f(x)$. Without loss of generality, we may assume that $y = \mathbf{0}$. We first demonstrate that $f(x)$ is bounded above near $\mathbf{0}$. Define the $n + 1$ points

$$\begin{aligned} v_{(0)} &= -\frac{r}{2n}\mathbf{1} \\ v_{(i)} &= re_{(i)} - \frac{r}{2n}\mathbf{1}, \quad 1 \leq i \leq n \end{aligned}$$

using the standard basis $e_{(i)}$ of R^n. It is easy to check that all of these points lie in $C(\mathbf{0}, r)$. Hence, any convex combination $\sum_{i=0}^n \alpha_i v_{(i)}$ also lies in $C(\mathbf{0}, r)$. Even more surprising, any point x in the open interval

$$J = \prod_{i=1}^n \Big(-\frac{r}{2n}, \frac{r}{2n}\Big)$$

can be represented as such a convex combination. This assertion follows from the component-by-component equation

$$x_i = r\Big(\alpha_i - \frac{1}{2n}\Big)$$

with $\alpha_i \in (0, 1/n)$ and the identity

$$\begin{aligned} x &= r\sum_{i=1}^{n} x_i e_{(i)} \\ &= r\sum_{i=1}^{n} \alpha_i e_{(i)} - \frac{r}{2n}\mathbf{1} \\ &= r\sum_{i=1}^{n} \alpha_i \Big(e_{(i)} - \frac{1}{2n}\mathbf{1}\Big) - \Big(1 - \sum_{i=1}^{n} \alpha_i\Big)\frac{r}{2n}\mathbf{1}. \end{aligned}$$

The boundedness of $f(x)$ on J now follows from the inequalities

$$\begin{aligned} f\Big(\sum_{i=0}^{n} \alpha_i v_{(i)}\Big) &\le \sum_{i=0}^{n} \alpha_i f(v_{(i)}) \\ &\le \max\{f(v_{(0)}), \ldots, f(v_{(n)})\}. \end{aligned}$$

Without affecting its convexity, we now rescale and translate $f(x)$ so that $f(\mathbf{0}) = 0$ and $f(x) \le 1$ on J. We also rescale x so that J contains the open ball $B(\mathbf{0}, 2)$. For any x in $B(\mathbf{0}, 2)$, we have

$$0 = f(\mathbf{0}) \le \frac{1}{2}f(x) + \frac{1}{2}f(-x).$$

It follows that $f(x)$ is bounded below by -1 on $B(\mathbf{0}, 2)$. The final step of the proof consists in choosing two distinct points x and z from the unit ball $B(\mathbf{0}, 1)$. If we define $w = z + t^{-1}(z - x)$ with $t = \|z - x\|$, then $w \in B(\mathbf{0}, 2)$,

$$z = \frac{t}{1+t}w + \frac{1}{1+t}x,$$

and

$$\begin{aligned} f(z) - f(x) &\le \frac{t}{1+t}f(w) + \frac{1}{1+t}f(x) - f(x) \\ &= \frac{t}{1+t}f(w) - \frac{t}{1+t}f(x) \\ &\le \frac{2t}{1+t} \\ &\le 2\|z - x\|. \end{aligned}$$

Switching the roles of x and z gives $|f(z) - f(x)| \le 2\|z - x\|$. This Lipschitz inequality establishes the continuity of $f(x)$ throughout $B(\mathbf{0}, 1)$. ■

Figure 5.1 illustrates how a tangent line to a convex curve lies below the curve. This property characterizes convex differentiable functions.

Proposition 5.3.2 *Let $f(x)$ be a differentiable function on the open convex set $S \subset \mathsf{R}^m$. Then $f(x)$ is convex if and only if*

$$f(y) \geq f(x) + df(x)(y - x) \tag{5.3}$$

for all $x, y \in S$. Furthermore, $f(x)$ is strictly convex if and only if strict inequality holds in inequality (5.3) when $y \neq x$.

Proof: If $f(x)$ is convex, then we can rearrange inequality (5.2) to give

$$\begin{aligned} \frac{f[x + (1-\alpha)(y-x)] - f(x)}{(1-\alpha)} &= \frac{f[\alpha x + (1-\alpha)y] - f(x)}{1-\alpha} \\ &\leq f(y) - f(x). \end{aligned}$$

Letting α tend to 1 proves inequality (5.3). To demonstrate the converse, let $z = \alpha x + (1-\alpha)y$. Then with obvious notational changes, inequality (5.3) implies

$$\begin{aligned} f(x) &\geq f(z) + df(z)(x - z) \\ f(y) &\geq f(z) + df(z)(y - z). \end{aligned}$$

Multiplying the first of these inequalities by α and the second by $1 - \alpha$ and adding the results produce

$$\alpha f(x) + (1-\alpha)f(y) \geq f(z) + df(z)(z - z) = f(z),$$

which is just inequality (5.2). The claims about strict convexity are left to the reader. ∎

One can relax the assumption of differentiability of Proposition 5.3.2 by introducing the notion of subdifferentiability. A function $f(x)$ defined on a convex set S possesses a subdifferential w^* at the point $y \in S$ provided $f(x) \geq f(y) + w^*(x - y)$ for all $x \in S$. The next proposition is one of the keys to the modern theory of convex functions.

Proposition 5.3.3 *A convex function $f(x)$ possesses a subdifferential at every point y of the interior of its domain S. If $f(x)$ is differentiable at y, then its unique subdifferential at that point is its differential.*

Proof: To generate a subdifferential at y, we separate the boundary point $(y^*, f[y])^*$ of the epigraph from the rest of the epigraph. Example 5.3.5 shows that the epigraph $\{(x, r) : x \in S, r \geq f(x)\}$ is a convex set. To prove that $(y^*, f[y])^*$ is a boundary point, we must exhibit a sequence of points from the complement of the epigraph that converges to $(y^*, f[y])^*$. The sequence $(y^*, f[y] - 1/k)^*$ qualifies. In fact, $(y^*, f[y])^*$ is also a boundary point of the closure of the epigraph. For this stronger statement, it suffices to show that $(y^*, f[y] - 1/k)^*$ is in the interior of the complement of the epigraph. By virtue of the continuity of $f(x)$ at y, there is a δ_k such that

$\|x-y\| < \delta_k$ implies $|f(x)-f(y)| < \frac{1}{2k}$. Any point $(x^*, r)^*$ with $\|x-y\| < \delta_k$ and $|r - f(y) + \frac{1}{k}| < \frac{1}{2k}$ satisfies

$$f(x) \ > \ f(y) - \frac{1}{2k} \ > \ r$$

and lies in the complement of the epigraph.

We are now in a position to invoke Proposition 5.2.2 and find a unit vector $(v^*, c)^*$ such that $v^*y + cf(y) \geq v^*x + cr$ for every point $(x^*, r)^*$ in the closure of the epigraph. The condition $c > 0$ is clearly impossible because we can take r arbitrarily large when $x = y$. The condition $c = 0$ is also impossible. Since y is an interior point of S, there exists a small ball centered at y and contained within S. The condition $v^*(y-x) \geq 0$ can hold for all points x in this ball only if $v = \mathbf{0}$. But this violates the fact that $(v^*, c)^*$ is a unit vector. Hence, the conclusion $c < 0$ is the only tenable one.

For the choice $r = f(x)$, we now divide the inequality

$$v^*y + cf(y) \ \geq \ v^*x + cf(x)$$

by c and rearrange. This yields

$$f(x) \ \geq \ f(y) + w^*(x - y)$$

and identifies $w^* = -c^{-1}v^*$ as a subdifferential.

For the final claim of the proposition, suppose w^* is a subdifferential different from $df(y)$. Taking limits in the inequality

$$\frac{f(y + tu) - f(y)}{t} \ \geq \ w^*u$$

as $t > 0$ tends to 0 implies $df(y)u \geq w^*u$ for all vectors u. The choice $u = w - \nabla f(y)$ produces the inequality $\|w - \nabla f(x)\|^2 \leq 0$ and compels the conclusion that $w^* = df(y)$. ■

Readers should note that our terminology is at variance with the literature on convexity, where subdifferentials are called subgradients and the set of subgradients is called the subdifferential. The classical definition is confusing given our convention that differentials and gradients are transposes of one another.

Example 5.3.6 *Subdifferential of $|x|$ on* R

For $x > 0$, the convex function $f(x) = |x|$ has differential $df(x) = 1$. This serves as its unique subdifferential. Likewise, the differential $df(x) = -1$ is appropriate when $x < 0$. At the nondifferentiable point 0, it is obvious geometrically that any number d with $|d| \leq 1$ is a subdifferential. ■

The existence of subdifferentials is tied to the existence of forward directional derivatives. In the case of a convex function $f(x)$ defined on an interval (a, b), we will prove the existence of one-sided derivatives by establishing the inequalities

$$\frac{f(y)-f(x)}{y-x} \leq \frac{f(z)-f(x)}{z-x} \leq \frac{f(z)-f(y)}{z-y} \tag{5.4}$$

for all points $x < y < z$ drawn from (a, b). If we write

$$y = \frac{z-y}{z-x}x + \frac{y-x}{z-x}z,$$

then both of these inequalities are rearrangements of the inequality

$$f(y) \leq \frac{z-y}{z-x}f(x) + \frac{y-x}{z-x}f(z).$$

Careful examination of the inequalities (5.4) with relabeling of points as necessary leads to the conclusion that the slope

$$\frac{f(y)-f(x)}{y-x}$$

is bounded below and increasing in y for x fixed. Similarly, this same slope is bounded above and increasing in x for y fixed. It follows that both one-sided derivatives exist at y and satisfy

$$\lim_{x \uparrow y} \frac{f(y)-f(x)}{y-x} \leq \lim_{z \downarrow y} \frac{f(z)-f(y)}{z-y}.$$

In view of the monotonicity properties of the slope, any number $df(y)$ between these two limits satisfies the requirements

$$\begin{aligned} f(x) &\geq f(y) + df(y)(x-y) \\ f(z) &\geq f(y) + df(y)(z-y) \end{aligned}$$

of a subdifferential.

This reasoning for convex functions defined on the real lines proves the existence of forward directional derivatives on higher-dimensional domains. Indeed, all one must do is define the function $g(t) = f(x+tv)$ of the real variable t for x and $y = x + v$ in the interior of the domain of $f(x)$. It is easy to check that the convexity of $f(x)$ carries over to $g(t)$. Furthermore,

$$\lim_{t \downarrow 0} \frac{f(x+tv)-f(x)}{t} = \lim_{t \downarrow 0} \frac{g(t)-g(0)}{t}.$$

It is useful to have simpler tests for convexity than inequalities (5.2) and (5.3). One such test involves the second differential $d^2 f(x)$ of a function $f(x)$.

Proposition 5.3.4 *Consider a twice differentiable function $f(x)$ on the open convex set $S \subset \mathsf{R}^m$. If its second differential $d^2 f(x)$ is positive semidefinite, then $f(x)$ is convex. If $d^2 f(x)$ is positive definite, then $f(x)$ is strictly convex.*

Proof: The expansion

$$\begin{aligned} f(y) &= f(x) + df(x)(y - x) \\ &\quad + (y - x)^* \int_0^1 d^2 f[x + t(y - x)](1 - t)\, dt\, (y - x) \end{aligned}$$

shows that

$$f(y) \geq f(x) + df(x)(y - x),$$

with strict inequality when $d^2 f(x)$ is positive definite for all x. ■

Example 5.3.7 *Strictly Convex Quadratic Functions*

If the matrix A is positive definite, then Proposition 5.3.4 implies that the quadratic function $f(x) = \frac{1}{2} x^t A x + b^t x + c$ is strictly convex. ■

Even Proposition 5.3.4 can be difficult to apply. The next proposition helps us to recognize convex functions by their closure properties.

Proposition 5.3.5 *Convex functions satisfy the following:*

(a) If $f(x)$ is convex and $g(x)$ is convex and increasing, then the functional composition $g \circ f(x)$ is convex.

(b) If $f(x)$ is convex, then the functional composition $f(Ax + b)$ of $f(x)$ with an affine function $Ax + b$ is convex.

(c) If $f(x)$ and $g(x)$ are convex and α and β are nonnegative constants, then $\alpha f(x) + \beta g(x)$ is convex.

(d) If $f(x)$ and $g(x)$ are convex, then $\max\{f(x), g(x)\}$ is convex.

(e) If $f_n(x)$ is a sequence of convex functions, then $\lim_{n \to \infty} f_n(x)$ is convex whenever it exists.

Proof: To prove assertion (a), we calculate

$$\begin{aligned} g \circ f[\alpha x + (1 - \alpha) y] &\leq g[\alpha f(x) + (1 - \alpha) f(y)] \\ &\leq \alpha g \circ f(x) + (1 - \alpha) g \circ f(y). \end{aligned}$$

The remaining assertions are left to the reader. ■

Part (a) of Proposition 5.3.5 implies that $e^{f(x)}$ is convex when $f(x)$ is convex and that $f(x)^\beta$ is convex when $f(x)$ is nonnegative and convex and $\beta > 1$. One case not covered by the Proposition is products. The counterexample $x^3 = x^2 x$ shows that the product of two convex functions is not necessarily convex. In some situations the limit of a sequence of convex functions is no longer finite. Many authors consider $+\infty$ to be a legitimate value for a convex function while $-\infty$ is illegitimate. For the sake of simplicity, we prefer to deal with functions having only finite values. In Chapter 11 we relax this restriction.

Example 5.3.8 *Differences of Convex Functions*

Although the class of convex functions is rather narrow, most well-behaved functions can be expressed as the difference of two convex functions. For example, consider a polynomial $p(x) = \sum_{m=0}^{n} p_m x^m$. The second derivative test shows that x^m is convex whenever m is even. If m is odd, then x^m is convex on $[0, \infty)$, and $-x^m$ is convex on $(-\infty, 0)$. Therefore,

$$x^m \quad = \quad \max\{x^m, 0\} - \max\{-x^m, 0\}$$

is the difference of two convex functions. Because the class of differences of convex functions is closed under the formation of linear combinations, it follows that $p(x)$ belongs to this larger class. ■

A positive function $f(x)$ is said to be log-convex if $\ln f(x)$ is convex. Log-convex functions have excellent closure properties as documented by the next proposition.

Proposition 5.3.6 *Log-convex functions satisfy the following:*

(a) If $f(x)$ is log-convex, then $f(x)$ is convex.

(b) If $f(x)$ is convex and $g(x)$ is log-convex and increasing, then the functional composition $g \circ f(x)$ is log-convex.

(c) If $f(x)$ is log-convex, then the functional composition $f(Ax + b)$ of $f(x)$ with an affine function $Ax + b$ is log-convex.

(d) If $f(x)$ is log-convex, then $f(x)^\alpha$ and $\alpha f(x)$ are log-convex for any $\alpha > 0$.

(e) If $f(x)$ and $g(x)$ are log-convex, then $f(x) + g(x)$, $f(x)g(x)$, and $\max\{f(x), g(x)\}$ are log-convex.

(f) If $f_n(x)$ is a sequence of log-convex functions, then $\lim_{n\to\infty} f_n(x)$ is log-convex whenever it exists and is positive.

Proof: Assertion (a) follows from part (a) of Proposition 5.3.5 after composing the functions e^x and $\ln f(x)$. To prove that the sum of log-convex functions is log-convex, we let $h(x) = f(x) + g(x)$ and apply Hölder's inequality as stated in Problem 13 of Chapter 4 and in Example 5.5.3 later in this chapter. Taking $\alpha = 1/p$ and $1-\alpha = 1/q$ consequently implies that

$$\begin{aligned} h[\alpha x+(1-\alpha)y] &= f[\alpha x+(1-\alpha)y]+g[\alpha x+(1-\alpha)y] \\ &\leq f(x)^\alpha f(y)^{1-\alpha}+g(x)^\alpha g(y)^{1-\alpha} \\ &\leq [f(x)+g(x)]^\alpha [f(y)+g(y)]^{1-\alpha} \\ &= h(x)^\alpha h(y)^{1-\alpha}. \end{aligned}$$

The remaining assertions are left to the reader. ■

Example 5.3.9 *The Convex Function of Gordon's Theorem*

In Proposition 4.3.2, we encountered the function

$$f(x) = \ln\left[\sum_{j=1}^{r} \exp(z_{(j)}^* x)\right].$$

Given the log-convexity of the functions $\exp(z_{(j)}^* x)$, we now recognize $f(x)$ as convex. This is one of the reasons for its success in Gordon's theorem. ■

Example 5.3.10 *Gamma Function*

Gauss's representation of the gamma function

$$\Gamma(z) = \lim_{n\to\infty} \frac{n! n^z}{z(z+1)\cdots(z+n)} \tag{5.5}$$

shows that it is log-convex on $(0,\infty)$ [63]. Indeed, one can easily check that n^z and $(z+k)^{-1}$ are log-convex and then apply the closure of the set of log-convex functions under the formation of products and limits. Note that invoking convexity in this argument is insufficient because the set of convex functions is not closed under the formation of products. Alternatively, one can deduce log-convexity from Euler's definition

$$\Gamma(z) = \int_0^\infty x^{z-1} e^{-x} dx$$

by viewing the integral as the limit of Riemann sums, each of which is log-convex. ■

5.4 Minimization of Convex Functions

Optimization theory is much simpler for convex functions than for ordinary functions. For instance, we have the following results:

Proposition 5.4.1 *Suppose that $f(y)$ is a convex function on the convex set $S \subset \mathsf{R}^m$. If x is a local minimum of $f(y)$, then it is a global minimum of $f(y)$, and the set $\{y \in S : f(y) = f(x)\}$ is convex. If $f(y)$ is strictly convex and x is a global minimum, then the set $\{y \in S : f(y) = f(x)\}$ consists of x alone.*

Proof: If $f(y) \leq f(x)$ and $f(z) \leq f(x)$, then

$$\begin{aligned} f[\alpha y + (1-\alpha)z] &\leq \alpha f(y) + (1-\alpha) f(z) \\ &\leq f(x) \end{aligned} \tag{5.6}$$

for any $\alpha \in [0,1]$. This shows that the set $\{y \in S : f(y) \leq f(x)\}$ is convex. Now suppose that $f(y) < f(x)$. Strict inequality then prevails between the extreme members of inequality (5.6) provided $\alpha > 0$. Taking $z = x$ and α close to 0 shows that x cannot serve as a local minimum. This contradiction demonstrates that x must be a global minimum. Finally, if $f(y)$ is strictly convex, then strict inequality holds in equality (5.6) for all $\alpha \in (0,1)$ and $z \neq y$. This leads to a contradiction when $z = x \neq y$ and $f(y) = f(x)$. ■

Example 5.4.1 *Piecewise Linear Functions*

The function $f(x) = |x|$ on the real line is piecewise linear. It attains its minimum of 0 at the point $x = 0$. The convex function $f(x) = \max\{1, |x|\}$ is also piecewise linear, but it attains its minimum throughout the interval $[-1,1]$. In both cases the set $\{y : f(y) = \min_x f(x)\}$ is convex. In higher dimensions, the convex function $f(x) = \max\{1, \|x\|\}$ attains its minimum of 1 throughout the closed ball $\|x\| \leq 1$. ■

Proposition 5.4.2 *Let $f(y)$ be a convex differentiable function on the convex set $S \subset \mathsf{R}^m$. If the point $x \in S$ satisfies*

$$df(x)(z - x) \geq 0$$

for every point $z \in S$, then x is a global minimum of $f(y)$. In particular, any stationary point of $f(y)$ is a global minimum. This result generalizes to nondifferentiable convex functions provided we view $df(x)$ as a subdifferential.

Proof: This assertion follows immediately from inequality (5.3) characterizing convex functions. ■

Example 5.4.2 *Minimum of y on $[0, \infty)$*

The convex function $f(y) = y$ has derivative $f'(y) = 1$. On the convex set $[0, \infty)$, we have $f'(0)(z - 0) = z \geq 0$ for any $z \in [0, \infty)$. Hence, 0 provides the minimum of y. Of course, this is consistent with the multiplier rule $f'(0) - 1 = 0$. ■

In Chapter 4 we found that the multiplier rule (4.1) is a necessary condition for a feasible point x to be a local minimum of the function $f(y)$ subject to the constraints

$$\begin{aligned} g_i(y) &= 0, \quad 1 \leq i \leq p \\ h_j(y) &\leq 0, \quad 1 \leq j \leq q. \end{aligned}$$

Proposition 4.5.1 states conditions under which the multiplier rule is also a sufficient condition. In the presence of convexity, we can do better than Proposition 4.5.1.

Proposition 5.4.3 *Suppose the functions $g_i(y)$ are affine and the functions $f(x)$ and $h_j(y)$ are convex. Then a feasible point x satisfying the multiplier rule (4.1) with $\lambda_0 = 1$ furnishes a global minimum of $f(x)$.*

Proof: Because the feasible region is convex, we can apply Proposition 5.4.2. If z is another feasible point, then the vector $v = z - x$ satisfies $0 = g_i(z) - g_i(x) = dg_i(x)v$ and

$$0 \geq h_j(z) - h_j(x) \geq dh_j(x)v \tag{5.7}$$

for each inequality constraint active at x. Taking the inner product of v with both sides of the multiplier equality

$$\nabla f(x) + \sum_{i=1}^{p} \lambda_i \nabla g_i(x) + \sum_{j=1}^{q} \mu_j \nabla h_j(x) = \mathbf{0}$$

therefore leads to

$$df(x)v = -\sum_{j=1}^{q} \mu_j dh_j(x)v \geq 0,$$

which is exactly the sufficient condition of Proposition 5.4.2. ■

Example 5.4.3 *Slater's Constraint Qualification*

In developing the multiplier rule for general objective functions $f(y)$, we imposed the Mangasarian-Fromovitz constraint qualification. We now show that this condition is implied by a simpler condition called the Slater constraint qualification given affine equality constraints and convex inequality constraints. Slater's condition postulates the existence of a feasible point z such that $h_j(z) < 0$ for all j. If x is a candidate minimum point and the row vectors $dg_i(x)$ are linearly independent, then the Mangasarian-Fromovitz constraint qualification involves finding a vector v with $dg_i(x)v = 0$ for all i and $dh_j(x)v < 0$ for all inequality constraints active at x. Since inequality (5.7) is strict at an active constraint for the vector $v = z - x$, the Mangasarian-Fromovitz constraint qualification follows. Nothing in this argument depends on the objective function $f(y)$ being convex. ■

Example 5.4.4 *Minimum of a Positive Definite Quadratic Function*

The quadratic function $f(x) = \frac{1}{2}x^*Ax + b^*x + c$ has differential

$$df(x) = x^*A + b^*$$

for A symmetric. Assuming that A is also invertible, the sole stationary point of $f(x)$ is $-A^{-1}b$. This point furnishes the minimum of $f(x)$ when A is positive definite. Given affine equality constraints and no inequality constraints, Proposition 5.4.3 shows that the candidate minimum point identified in Example 4.2.5 furnishes the global minimum of $f(x)$. ■

Example 5.4.5 *Maximum Likelihood for the Multivariate Normal*

The sample mean and sample variance

$$\begin{aligned}
\bar{y} &= \frac{1}{k}\sum_{j=1}^{k} y_j \\
S &= \frac{1}{k}\sum_{j=1}^{k}(y_j - \bar{y})^*(y_j - \bar{y})
\end{aligned}$$

are also the maximum likelihood estimates of the theoretical mean μ and theoretical variance Ω of a random sample $y_1, \ldots, y_k$ from a multivariate normal. (See the Appendix for a review of the multivariate normal.) To prove this fact, we first note that maximizing the loglikelihood function

$$\begin{aligned}
& -\frac{k}{2}\ln\det\Omega - \frac{1}{2}\sum_{j=1}^{k}(y_j - \mu)^*\Omega^{-1}(y_j - \mu) \\
= \; & -\frac{k}{2}\ln\det\Omega - \frac{k}{2}\mu^*\Omega^{-1}\mu + \Big(\sum_{j=1}^{k} y_j\Big)^*\Omega^{-1}\mu - \frac{1}{2}\sum_{j=1}^{k} y_j^*\Omega^{-1}y_j \\
= \; & -\frac{k}{2}\ln\det\Omega - \frac{1}{2}\operatorname{tr}\Big[\Omega^{-1}\sum_{j=1}^{k}(y_j - \mu)(y_j - \mu)^*\Big]
\end{aligned}$$

constitutes a special case of the previous example with $A = k\Omega^{-1}$ and $b = -\Omega^{-1}\sum_{j=1}^{k} y_j$. This leads to the same estimate $\hat{\mu} = \bar{y}$ regardless of the value of Ω.

To estimate Ω, we exploit the Cholesky decompositions $\Omega = LL^*$ and $S = MM^*$ under the assumption that both Ω and S are invertible. (See Problems 15 and 16 for a development of the Cholesky decomposition of a positive definite matrix.) In view of the identities $\Omega^{-1} = (L^{-1})^*L^{-1}$ and $\det\Omega = (\det L)^2$, the loglikelihood becomes

$$\begin{aligned}
& k\ln\det L^{-1} - \frac{k}{2}\operatorname{tr}\Big[(L^{-1})^*L^{-1}MM^*\Big] \\
= \; & k\ln\det\big(L^{-1}M\big) - \frac{k}{2}\operatorname{tr}\Big[(L^{-1}M)(L^{-1}M)^*\Big] - k\ln\det M
\end{aligned}$$

using the cyclic permutation property of the matrix trace function. Because products and inverses of lower triangular matrices are lower triangular, the matrix $R = L^{-1}M$ ranges over the set of lower triangular matrices with positive diagonal entries as L ranges over the same set. This permits us to reparameterize and estimate $R = (r_{ij})$ instead of L. Up to an irrelevant additive constant, the loglikelihood reduces to

$$k \ln \det R - \frac{k}{2} \operatorname{tr}(RR^*) = k \sum_i \ln r_{ii} - \frac{k}{2} \sum_i \sum_{j=1}^{i} r_{ij}^2.$$

Clearly, this is maximized by taking $r_{ij} = 0$ for $j \neq i$. Differentiation of the concave function $k \ln r_{ii} - \frac{k}{2} r_{ii}^2$ shows that it is maximized by taking $r_{ii} = 1$. In other words, the maximum likelihood estimator $\hat{R}$ is the identity matrix I. This implies that $\hat{L} = M$ and consequently that $\hat{\Omega} = S$. ■

Example 5.4.6 *Geometric Programming*

The function

$$f(t) = \sum_{i=1}^{j} c_i \prod_{k=1}^{n} t_k^{\beta_{ik}}$$

is called a posynomial if all components $t_1, \ldots, t_n$ of the argument t and all coefficients $c_1, \ldots, c_j$ are positive. The powers β_{ik} may be positive or negative. For instance, $t_1^{-1} + 2t_1^3 t_2^{-2}$ is a posynomial on R^2. Geometric programming deals with the minimization of a posynomial $f(t)$ subject to posynomial inequality constraints of the form $h_j(t) \leq 1$ for $1 \leq j \leq q$.

Better understanding of geometric programming can be achieved by making the change of variables $t_k = e^{x_k}$. This eliminates the constraint $t_k > 0$ and shows that

$$g(x) = \sum_{i=1}^{j} c_i \prod_{k=1}^{n} t_k^{\beta_{ik}} = \sum_{i=1}^{j} c_i e^{\beta_{(i)}^* x}$$

is log-convex in the transformed parameters. The reparameterized constraint functions are likewise log-convex and define a convex feasible region S. If the vectors $\beta_{(1)}, \ldots, \beta_{(j)}$ span R^n, then the expression

$$d^2 g(x) = \sum_{i=1}^{j} c_i e^{\beta_{(i)}^* x} \beta_{(i)} \beta_{(i)}^*$$

for the second differential proves that $g(x)$ is strictly convex. It follows that if $g(x)$ possesses a minimum, then it is achieved at a single point. ■

Example 5.4.7 *Quasi-Convexity*

If $f(x)$ is convex and $g(z)$ is an increasing function of the real variable z, then the inequality

$$\begin{aligned} f[\alpha x + (1-\alpha)y] &\leq \alpha f(x) + (1-\alpha)f(y) \\ &\leq \max\{f(x), f(y)\} \end{aligned}$$

implies that the function $h(x) = g \circ f(x)$ satisfies

$$h[\alpha x + (1-\alpha)y] \leq \max\{h(x), h(y)\}. \qquad (5.8)$$

This inequality is taken as the definition of quasi-convexity for any function $h(x)$. If $f(x)$ is strictly convex and $g(z)$ is strictly increasing, then strict inequality prevails in inequality (5.8) when $\alpha \in (0,1)$ and $y \neq x$. Once again this implication can be turned into a definition. The importance of strict quasi-convexity lies in the fact that a strictly quasi-convex function possesses at most one local minimum, and if a local minimum exists, then it is necessarily the global minimum.

Similar considerations apply to concave and quasi-concave functions. For example, the function $h(x) = e^{-(x-\mu)^2}$ is strictly quasi-concave because it is the composition of the strictly increasing function $g(y) = e^y$ with the strictly concave function $f(x) = -(x-\mu)^2$. It is clear that $h(x)$ has a global maximum at $x = \mu$. ■

5.5 Moment Inequalities

In this section we assume that readers have a good grasp of probability theory. For those with limited background, most of the material can be comprehended by restricting attention to discrete random variables.

Inequalities give important information about the magnitude of probabilities and expectations without requiring their exact calculation. The Cauchy-Schwarz inequality $|\,\mathrm{E}(XY)| \leq \mathrm{E}(X^2)^{1/2}\,\mathrm{E}(Y^2)^{1/2}$ is one of the most useful of the classical inequalities. (The reader should check that this is just a disguised form of the Cauchy-Schwarz inequality of Chapter 1.) It is also one of the easiest to remember because it is equivalent to the fact that a correlation coefficient must lie on the interval $[-1,1]$. Equality occurs in the Cauchy-Schwarz inequality if and only if X is proportional to Y or vice versa.

Markov's inequality is another widely applied bound. Let $g(x)$ be a nonnegative, increasing function, and let X be a random variable such that $g(X)$ has finite expectation. Then Markov's inequality

$$\Pr(X \geq c) \leq \frac{\mathrm{E}[g(X)]}{g(c)}$$

holds for any constant c for which $g(c) > 0$ and follows logically by taking expectations in the inequality $g(c)1_{\{X \geq c\}} \leq g(X)$. Chebyshev's inequality is the special case of Markov's inequality with $g(x) = x^2$ applied to the random variable $X - \mathrm{E}(X)$. This inequality reads

$$\Pr[|X - \mathrm{E}(X)| \geq c] \quad \leq \quad \frac{\mathrm{Var}(X)}{c^2}.$$

In large deviation theory, we take $g(x) = e^{tx}$ and $c > 0$ and choose $t > 0$ to minimize the right-hand side of the inequality $\Pr(X \geq c) \leq e^{-ct}\,\mathrm{E}(e^{tX})$ involving the moment generating function of X. Problem 20 provides a typical example of this Chernoff bound.

Our next example involves a nontrivial application of Chebyshev's inequality. In preparation for the example, we recall that a binomially distributed random variable S_n has distribution $\Pr(S_n = k) = \binom{n}{k} x^k (1-x)^{n-k}$. Here S_n is interpreted as the number of successes in n independent trials with success probability x per trial [39]. The mean and variance of S_n are $\mathrm{E}(S_n) = nx$ and $\mathrm{Var}(S_n) = nx(1-x)$.

Example 5.5.1 *Weierstrass's Approximation Theorem*

Weierstrass showed that a continuous function $f(x)$ on $[0, 1]$ can be uniformly approximated to any desired degree of accuracy by a polynomial. Bernstein's lovely proof of this fact relies on applying Chebyshev's inequality to the random variable S_n/n derived from the binomial random variable S_n just discussed. The corresponding candidate polynomial is defined by the expectation

$$\mathrm{E}\left[f\left(\frac{S_n}{n}\right)\right] \quad = \quad \sum_{k=0}^{n} f\left(\frac{k}{n}\right)\binom{n}{k} x^k (1-x)^{n-k}.$$

Note that $\mathrm{E}(S_n/n) = x$ and

$$\mathrm{Var}\left(\frac{S_n}{n}\right) \quad = \quad \frac{x(1-x)}{n} \quad \leq \quad \frac{1}{4n}.$$

Now given an arbitrary $\epsilon > 0$, one can find by the uniform continuity of $f(x)$ a $\delta > 0$ such that $|f(u) - f(v)| < \epsilon$ whenever $|u - v| < \delta$. If $\|f\|_\infty = \sup |f(x)|$ on $[0, 1]$, then Chebyshev's inequality implies

$$\begin{aligned}
&\left|\mathrm{E}\left[f\left(\frac{S_n}{n}\right)\right] - f(x)\right| \\
&\leq \quad \mathrm{E}\left[\left|f\left(\frac{S_n}{n}\right) - f(x)\right|\right] \\
&\leq \quad \epsilon \Pr\left(\left|\frac{S_n}{n} - x\right| < \delta\right) + 2\|f\|_\infty \Pr\left(\left|\frac{S_n}{n} - x\right| \geq \delta\right) \\
&\leq \quad \epsilon + \frac{2\|f\|_\infty x(1-x)}{n\delta^2} \\
&\leq \quad \epsilon + \frac{\|f\|_\infty}{2n\delta^2}.
\end{aligned}$$

Taking $n \geq \|f\|_\infty/(2\epsilon\delta^2)$ then gives $\left|\mathrm{E}\left[f\left(\frac{S_n}{n}\right)\right] - f(x)\right| \leq 2\epsilon$ regardless of the chosen $x \in [0,1]$. ■

Proposition 5.5.1 (Jensen's Inequality) *Let the values of the random variable W be confined to the possibly infinite interval (a,b). If $h(w)$ is convex on (a,b), then $\mathrm{E}[h(W)] \geq h[\mathrm{E}(W)]$, provided both expectations exist. For a strictly convex function $h(w)$, equality holds in Jensen's inequality if and only if $W = \mathrm{E}(W)$ almost surely.*

Proof: Let $dh(v)$ denote a subdifferential of $h(w)$ at the point $v = \mathrm{E}(W)$. Then Jensen's inequality follows from Proposition 5.3.2 after taking expectations in the inequality

$$h(W) \quad \geq \quad h(v) + dh(v)(W - v). \tag{5.9}$$

If $h(w)$ is strictly convex, and W is not constant, then inequality (5.9) is strict with positive probability. Hence, strict inequality prevails in Jensen's inequality. ■

Jensen's inequality is the key to a host of other inequalities. Here is one important example.

Example 5.5.2 *Schlömilch's Inequality for Weighted Means*

If X is a positive random variable, then we define the weighted mean function $\mathrm{M}(p) = \mathrm{E}(X^p)^{\frac{1}{p}}$. For the sake of argument, we assume that $\mathrm{M}(p)$ exists and is finite for all real p. Typical values of $\mathrm{M}(p)$ are $\mathrm{M}(1) = \mathrm{E}(X)$ and $\mathrm{M}(-1) = 1/\mathrm{E}(X^{-1})$. To make $\mathrm{M}(p)$ continuous at $p = 0$, it turns out that we should set $\mathrm{M}(0) = e^{\mathrm{E}(\ln X)}$. The reader is asked to check this fact in Problem 26. Here we are more concerned with proving Schlömilch's assertion that $\mathrm{M}(p)$ is an increasing function of p. To simplify our proof, let us assume that $a \leq X \leq b$ for positive constants a and b. This condition permits us to differentiate under the expectation sign in

$$\begin{aligned} \frac{d}{dp} \ln \mathrm{M}(p) &= \frac{d}{dp}\left[\frac{1}{p} \ln \mathrm{E}(e^{p \ln X})\right] \\ &= -\frac{1}{p^2} \ln \mathrm{E}(X^p) + \frac{1}{p^2 \, \mathrm{E}(X^p)} p \, \mathrm{E}(X^p \ln X) \\ &= \frac{\mathrm{E}(X^p \ln X^p) - \mathrm{E}(X^p) \ln \mathrm{E}(X^p)}{p^2 \, \mathrm{E}(X^p)}. \end{aligned}$$

Because $f(u) = u \ln u$ is a convex function, Jensen's inequality implies that $\mathrm{E}(X^p \ln X^p) - \mathrm{E}(X^p) \ln \mathrm{E}(X^p) \geq 0$. Hence, $\mathrm{M}(p)$ has a nonnegative derivative and is increasing to both the left and right of 0. Continuity of $\mathrm{M}(p)$ at $p = 0$ then completes the proof that $\mathrm{M}(p)$ is increasing.

For readers familiar with Lebesgue's dominated convergence theorem, we can relax the boundedness condition on X by replacing X by

$$X_n \quad = \quad \frac{1}{n} 1_{\{X < \frac{1}{n}\}} + X 1_{\{\frac{1}{n} \leq X \leq n\}} + n 1_{\{X > n\}}$$

and take limits in the valid inequality $\mathrm{E}(X_n^p)^{\frac{1}{p}} \leq \mathrm{E}(X_n^q)^{\frac{1}{q}}$. Note that the dominated convergence theorem with dominating functions $X_n^p \leq 1 + X^p$ and $X_n^q \leq 1 + X^q$ justifies this operation for $p, q \neq 0$. If either $p = 0$ or $q = 0$, then Schlömilch's claim follows from dividing the obvious Jensen's inequality $\mathrm{E}(\ln X^r) \leq \ln \mathrm{E}(X^r)$ by r and exponentiating the result.

When the random variable X is confined to the space $\{1, \ldots, n\}$ equipped with the uniform probabilities $p_i = 1/n$, Schlömilch's inequalities for the values $p = -1$, 0, and 1 reduce to the classical inequalities

$$\frac{1}{\frac{1}{n}\left(\frac{1}{x_1} + \cdots \frac{1}{x_n}\right)} \leq \left(x_1 \cdots x_n\right)^{\frac{1}{n}} \leq \frac{1}{n}\left(x_1 + \cdots + x_n\right)$$

relating the harmonic, geometric, and arithmetic means. ■

Example 5.5.3 *Hölder's Inequality*

Consider two random variables X and Y and two numbers $p > 1$ and $q > 1$ such that $p^{-1} + q^{-1} = 1$. Then Hölder's inequality

$$|\mathrm{E}(XY)| \leq \mathrm{E}(|X|^p)^{\frac{1}{p}} \, \mathrm{E}(|Y|^q)^{\frac{1}{q}} \tag{5.10}$$

generalizes the Cauchy-Schwarz inequality whenever the indicated expectations on its right exist. To prove (5.10), it clearly suffices to assume that X and Y are nonnegative. It also suffices to take $\mathrm{E}(X^p) = \mathrm{E}(Y^q) = 1$ once we divide the left-hand side of (5.10) by its right-hand side. Now set $\alpha = p^{-1}$, and let Z be a random variable equal to $u \geq 0$ with probability α and equal to $v \geq 0$ with probability $1 - \alpha$. Schlömilch's inequality $M(0) \leq M(1)$ for Z says

$$u^\alpha v^{1-\alpha} \leq \alpha u + (1 - \alpha) v. \tag{5.11}$$

If we substitute X^p for u and Y^q for v in this inequality and take expectations, then we find that $\mathrm{E}(XY) \leq \alpha + 1 - \alpha = 1$ as required.

We can avoid the application of Schlömilch's inequality in this argument by proving inequality (5.11) directly. Because the function $-z^\alpha$ is convex, inequality (5.3) entails

$$z^\alpha - 1 \leq \alpha(z - 1)$$

for $z > 0$. If we put $z = u/v$, then the preceding inequality yields (5.11). Of course, if either u or v is 0, then inequality (5.11) is trivial. ■

5.6 Problems

1. On which intervals are the following functions convex: e^x, e^{-x}, x^n for n an integer, $|x|^p$ for $p \geq 1$, $\sqrt{1 + x^2}$, $x \ln x$, and $\cosh x$? On these intervals, which functions are log-convex?

2. Demonstrate that the function $f(x) = x^n - na \ln x$ is convex on $(0, \infty)$ for any positive real number a and nonnegative integer n. Where does its minimum occur?

3. Show that Riemann's zeta function

$$\zeta(s) = \sum_{n=1}^{\infty} \frac{1}{n^s}$$

is log-convex for $s > 1$.

4. Show that the function

$$f(x) = \begin{cases} \frac{1-e^{-xt}}{x} & x \neq 0 \\ t & x = 0 \end{cases}$$

is log-convex for $t > 0$ fixed. (Hints: Use either the second derivative test, or express $f(x)$ as the integral

$$f(x) = \int_0^t e^{-xs} ds,$$

and use the closure properties of log-convex functions.)

5. Use Proposition 5.3.2 to prove the Cauchy-Schwarz inequality.

6. Prove the strict convexity assertions of Proposition 5.3.2.

7. Prove parts (b), (c), and (d) of Proposition 5.3.5.

8. Prove the unproved assertions of Proposition 5.3.6.

9. Let $f(x)$ be a continuous function on the real line satisfying

$$f\left[\frac{1}{2}(x+y)\right] \leq \frac{1}{2}f(x) + \frac{1}{2}f(y).$$

Demonstrate that $f(x)$ is convex.

10. If $f(x)$ is a nondecreasing function on the interval $[a, b]$, then show that $g(x) = \int_a^x f(y)dy$ is a convex function on $[a, b]$.

11. The Bohr-Mollerup theorem asserts that $\Gamma(z)$ is the only log-convex function on the interval $(0, \infty)$ that satisfies $\Gamma(1) = 1$ and the factorial identity $\Gamma(z + 1) = z\Gamma(z)$ for all z. We have seen that $\Gamma(z)$ has these properties. Prove conversely that any function $G(z)$ with these properties coincides with $\Gamma(z)$. (Hints: Check the inequalities

$$\begin{aligned} G(n+z) &\leq G(n)^{1-z}G(n)^z n^z = (n-1)!n^z \\ G(n+1) &\leq G(n+z)^z G(n+1+z)^{1-z} = G(n+z)(n+z)^{1-z} \end{aligned}$$

for all positive integers n and real numbers $z \in (0,1)$. These in turn yield the inequalities

$$\frac{n!n^z}{z(z+1)\cdots(z+n)} \leq G(z) \leq \frac{n!n^z}{z(z+1)\cdots(z+n)} \cdot \frac{z+n}{n}.$$

Taking limits on n shows that $G(z)$ equals Gauss's infinite product expansion of $\Gamma(z)$. Note that this proof simultaneously validates Gauss's expansion (5.5).)

12. Let $f(x)$ be a real-valued differentiable function on R^n. If $f(x)$ is strictly convex, prove that $df(x) = df(y)$ if and only if $x = y$.

13. Let $f(x)$ be a convex differentiable function on R^n. Show that the function

$$g(x) = f(x) + \epsilon\|x\|^2$$

is strictly convex for $\epsilon > 0$ and that the set $\{x \in \mathsf{R}^n : g(x) \leq g(y)\}$ is compact for any y.

14. Find the minimum of the function

$$g(x_1, x_2) = 2x_1^2 + x_2^2 + \frac{1}{2x_1^2 + x_2^2}$$

on R^2. (Hint: Consider $f(x) = x + 1/x$ on R.)

15. Let A be an $n \times n$ positive definite matrix. The Cholesky decomposition B of A is a lower-triangular matrix with positive diagonal entries such that $A = BB^*$. To prove that such a decomposition exists we can argue by induction. Why is the case of a 1×1 matrix trivial? Now suppose A is partitioned as

$$A = \begin{pmatrix} A_{11} & A_{12} \\ A_{21} & A_{22} \end{pmatrix}.$$

Applying the induction hypothesis, there exist matrices C_{11} and D_{22} such that

$$\begin{aligned} C_{11}C_{11}^* &= A_{11} \\ D_{22}D_{22}^* &= A_{22} - A_{21}A_{11}^{-1}A_{12}. \end{aligned}$$

Prove that

$$B = \begin{pmatrix} C_{11} & \mathbf{0} \\ A_{21}(C_{11}^*)^{-1} & D_{22} \end{pmatrix}$$

gives the desired decomposition. Extend this argument to show that B is uniquely determined.

16. Continuing Problem 15, show that one can compute the Cholesky decomposition $B = (b_{ij})$ of $A = (a_{ij})$ by the recurrence relations

$$\begin{aligned} b_{jj} &= \sqrt{a_{jj} - \sum_{k=1}^{j-1} b_{jk}^2} \\ b_{ij} &= \frac{a_{ij} - \sum_{k=1}^{j-1} b_{ik} b_{jk}}{b_{jj}}, \quad i > j \end{aligned}$$

for columns $j = 1$, $j = 2$, and so forth until column $j = n$. How can you compute $\det A$ in terms of the entries of B?

17. Prove that the set of lower triangular matrices with positive diagonal entries is closed under matrix multiplication and matrix inversion.

18. If the random variable X has values in the interval $[a, b]$, then show that $\mathrm{Var}(X) \leq (b-a)^2/4$. (Hints: Reduce to the case $[a, b] = [0, 1]$. If $\mathrm{E}(X) = p$, then show that $\mathrm{Var}(X) \leq p(1-p)$.)

19. Suppose $g(x)$ is a function such that $g(x) \leq 1$ for all x and $g(x) \leq 0$ for $x \leq c$. Demonstrate the inequality

$$\Pr(X \geq c) \geq \mathrm{E}[g(X)] \tag{5.12}$$

for any random variable X [39]. Verify that the polynomial

$$g(x) = \frac{(x-c)(c+2d-x)}{d^2}$$

with $d > 0$ satisfies the stated conditions leading to inequality (5.12). If X is nonnegative with $\mathrm{E}(X) = 1$ and $\mathrm{E}(X^2) = \beta$ and $c \in (0, 1)$, then prove that the choice $d = \beta/(1-c)$ yields

$$\Pr(X \geq c) \geq \frac{(1-c)^2}{\beta}.$$

Finally, if $\mathrm{E}(X^2) = 1$ and $\mathrm{E}(X^4) = \beta$, show that

$$\Pr(|X| \geq c) \geq \frac{(1-c^2)^2}{\beta}.$$

20. Let X be a Poisson random variable with mean λ. Demonstrate that the Chernoff bound

$$\Pr(X \geq c) \leq \inf_{t>0} e^{-ct}\, \mathrm{E}(e^{tX})$$

amounts to

$$\Pr(X \geq c) \leq \frac{(\lambda e)^c}{c^c} e^{-\lambda}$$

for any integer $c > \lambda$. Note that $\Pr(X = i) = \lambda^i e^{-\lambda}/i!$ for all nonnegative integers i.

21. Use Jensen's inequality to prove the inequality

$$\prod_{k=1}^{n} x_k^{\alpha_k} + \prod_{k=1}^{n} y_k^{\alpha_k} \quad \leq \quad \prod_{k=1}^{n} (x_k + y_k)^{\alpha_k}$$

for positive numbers x_k and y_k and nonnegative numbers α_k with sum $\sum_{k=1}^{n} \alpha_k = 1$. Prove the inequality

$$\left(1 + \prod_{k=1}^{n} x_k^{\alpha_k}\right)^{-1} \quad \leq \quad \sum_{k=1}^{n} \frac{\alpha_k}{1 + x_k}$$

when all $x_k \geq 1$ and the reverse inequality when all $x_k \in (0, 1]$.

22. Let $B_n f(x) = \mathrm{E}[f(S_n/n)]$ denote the Bernstein polynomial of degree n approximating $f(x)$ as discussed in Example 5.5.1. Prove that

    (a) $B_n f(x)$ is linear in $f(x)$,
    (b) $B_n f(x) \geq 0$ if $f(x) \geq 0$,
    (c) $B_n f(x) = f(x)$ if $f(x)$ is linear,
    (d) $B_n x(1 - x) = \frac{n-1}{n} x(1 - x)$.

23. Suppose the function $f(x)$ has continuous derivative $f'(x)$. For $\delta > 0$ show that Bernstein's polynomial satisfies the bound

$$\left| \mathrm{E}\left[f\left(\frac{S_n}{n}\right)\right] - f(x) \right| \quad \leq \quad \delta \|f'\|_\infty + \frac{\|f\|_\infty}{2n\delta^2}.$$

Conclude from this estimate that $\left\| \mathrm{E}\left[f\left(\frac{S_n}{n}\right)\right] - f \right\|_\infty = O(n^{-\frac{1}{3}})$.

24. Let $f(x)$ be a convex function on $[0, 1]$. Prove that the Bernstein polynomial of degree n approximating $f(x)$ is also convex. (Hint: Show that

$$\begin{aligned} \frac{d^2}{dx^2} \mathrm{E}\left[f\left(\frac{S_n}{n}\right)\right] \quad &= \quad n(n-1)\Big\{ \mathrm{E}\left[f\left(\frac{S_{n-2}+2}{n}\right)\right] \\ &\quad -2\, \mathrm{E}\left[f\left(\frac{S_{n-2}+1}{n}\right)\right] + \mathrm{E}\left[f\left(\frac{S_{n-2}}{n}\right)\right]\Big\}. \end{aligned}$$

in the notation of Example 5.5.1.)

25. Suppose $1 \leq p < \infty$. For a random variable X with $\mathrm{E}(|X|^p) < \infty$, define the norm $\|X\|_p = \mathrm{E}(X^p)^{\frac{1}{p}}$. Now prove Minkowski's triangle inequality $\|X+Y\|_p \leq \|X\|_p + \|Y\|_p$. (Hint: Apply Hölder's inequality to the right-hand side of

$$\mathrm{E}(|X + Y|^p) \quad \leq \quad \mathrm{E}(|X| \cdot |X + Y|^{p-1}) + \mathrm{E}(|Y| \cdot |X + Y|^{p-1})$$

and rearrange the result.

26. Suppose X is a random variable satisfying $0 < a \leq X \leq b < \infty$. Use L'Hôpital's rule to prove that the weighted mean $\mathrm{M}(p) = \mathrm{E}(X^p)^{\frac{1}{p}}$ is continuous at $p = 0$ if we define $\mathrm{M}(0) = e^{\mathrm{E}(\ln X)}$.

6

The MM Algorithm

6.1 Introduction

Most practical optimization problems defy exact solution. In the current chapter we discuss an optimization method that relies heavily on convexity arguments and is particularly useful in high-dimensional problems such as image reconstruction [86]. This iterative method is called the MM algorithm. One of the virtues of this acronym is that it does double duty. In minimization problems, the first M of MM stands for majorize and the second M for minimize. In maximization problems, the first M stands for minorize and the second M for maximize. When it is successful, the MM algorithm substitutes a simple optimization problem for a difficult optimization problem. Simplicity can be attained by (a) avoiding large matrix inversions, (b) linearizing an optimization problem, (c) separating the variables of an optimization problem, (d) dealing with equality and inequality constraints gracefully, and (e) turning a nondifferentiable problem into a smooth problem. In simplifying the original problem, we must pay the price of iteration or iteration with a slower rate of convergence.

Statisticians have vigorously developed a special case of the MM algorithm called the EM algorithm, which revolves around notions of missing data [25, 82, 93]. We present the EM algorithm in the next chapter. We prefer to present the MM algorithm first because of its greater generality, its more obvious connection to convexity, and its weaker reliance on difficult statistical principles.

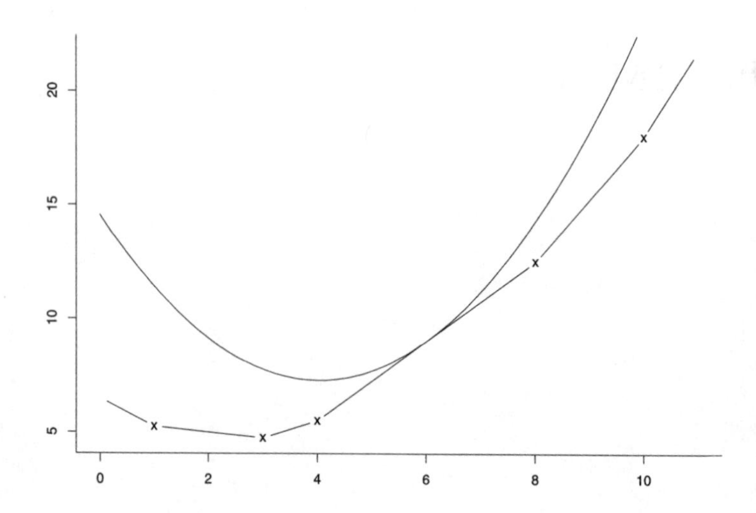

FIGURE 6.1. A Quadratic Majorizing Function for the Piecewise Linear Function $f(x) = |x-1| + |x-3| + |x-4| + |x-8| + |x-10|$ at the Point $x_{(m)} = 6$

6.2 Philosophy of the MM Algorithm

A function $g(x \mid x_{(m)})$ is said to majorize a function $f(x)$ at $x_{(m)}$ provided

$$\begin{aligned} f(x_{(m)}) &= g(x_{(m)} \mid x_{(m)}) \\ f(x) &\leq g(x \mid x_{(m)}), \qquad x \neq x_{(m)}. \end{aligned} \tag{6.1}$$

In other words, the surface $x \mapsto g(x \mid x_{(m)})$ lies above the surface $f(x)$ and is tangent to it at the point $x = x_{(m)}$. Here $x_{(m)}$ represents the current iterate in a search of the surface $f(x)$. Figure 6.1 provides a simple one-dimensional example.

In the minimization version of the MM algorithm, we minimize the surrogate majorizing function $g(x \mid x_{(m)})$ rather than the actual function $f(x)$. If $x_{(m+1)}$ denotes the minimum of the surrogate $g(x \mid x_{(m)})$, then we can show that the MM procedure forces $f(x)$ downhill. Indeed, the inequality

$$\begin{aligned} f(x_{(m+1)}) &= g(x_{(m+1)} \mid x_{(m)}) + f(x_{(m+1)}) - g(x_{(m+1)} \mid x_{(m)}) \\ &\leq g(x_{(m)} \mid x_{(m)}) + f(x_{(m)}) - g(x_{(m)} \mid x_{(m)}) \\ &= f(x_{(m)}) \end{aligned} \tag{6.2}$$

follows directly from the fact $g(x_{(m+1)} \mid x_{(m)}) \leq g(x_{(m)} \mid x_{(m)})$ and definition (6.1). The descent property (6.2) lends the MM algorithm remarkable numerical stability. Strictly speaking, it depends only on decreasing

$g(x \mid x_{(m)})$, not on minimizing $g(x \mid x_{(m)})$. This fact has practical consequences when the minimum of $g(x \mid x_{(m)})$ cannot be found exactly. When $f(x)$ is strictly convex, one can show with a few additional mild hypotheses that the iterates $x_{(m)}$ converge to the global minimum of $f(x)$ regardless of the initial point $x_{(0)}$.

Problem 1 makes the point that the majorization relation between functions is closed under the formation of sums, nonnegative products, limits, and composition with an increasing function. These rules permit us to work piecemeal in simplifying complicated objective functions. With obvious changes, the MM algorithm also applies to maximization rather than to minimization. To maximize a function $f(x)$, we minorize it by a surrogate function $g(x \mid x_{(m)})$ and maximize $g(x \mid x_{(m)})$ to produce the next iterate $x_{(m+1)}$.

The reader might well object that the MM algorithm is not so much an algorithm as a vague philosophy for deriving an algorithm. The same objection applies to the EM algorithm. As we proceed through the current chapter, we hope the various examples will convince the reader of the value of a unifying principle and a framework for attacking concrete problems. The strong connection of the MM algorithm to convexity and inequalities has the natural pedagogical advantage of building on the material presented in previous chapters.

6.3 Majorization and Minorization

We will feature four methods for constructing majorizing functions. Two of these simply adapt the inequality

$$f\Big(\sum_i \alpha_i t_i\Big) \;\le\; \sum_i \alpha_i f(t_i)$$

defining a convex function $f(t)$. It is easy to identify convex functions on the real line, so the first method composes such a function with a linear function c^*x to create a new convex function of the vector x. Invoking the definition of convexity with $\alpha_i = c_i y_i / c^* y$ and $t_i = c^* y \, x_i / y_i$ then yields

$$\begin{aligned} f(c^*x) &\le \sum_i \frac{c_i y_i}{c^* y} f\Big(\frac{c^* y}{y_i} x_i\Big) \qquad (6.3) \\ &= g(x \mid y), \end{aligned}$$

provided all of the components of the vectors c, x, and y are positive. The surrogate function $g(x \mid y)$ equals $f(c^*y)$ when $x = y$. One of the virtues of applying inequality (6.3) in defining a surrogate function is that it separates parameters in the surrogate function. This feature is critically important in high-dimensional problems because it reduces optimization over x to a sequence of one-dimensional optimizations over each component x_i. The argument establishing inequality (6.3) is equally valid if we replace the parameter vector x throughout by a vector-valued function $h(x)$ of x. The genetics problem in the next section illustrates this variant of the technique.

To relax the positivity restrictions on the vectors c, x, and y, De Pierro [27] suggested in a medical imaging context the alternative majorization

$$\begin{aligned} f(c^*x) &\leq \sum_i \alpha_i f\left\{\frac{c_i}{\alpha_i}(x_i - y_i) + c^*y\right\} \quad (6.4) \\ &= g(x \mid y) \end{aligned}$$

for a convex function $f(t)$. Here all $\alpha_i \geq 0$, $\sum_i \alpha_i = 1$, and $\alpha_i > 0$ whenever $c_i \neq 0$. In practice, we must somehow tailor the α_i to the problem at hand. Among the obvious candidates for the α_i are

$$\alpha_i = \frac{|c_i|^p}{\sum_j |c_j|^p}$$

for $p \geq 0$. When $p = 0$, we interpret α_i as 0 if $c_i = 0$ and as $1/q$ if c_i is one among q nonzero coefficients.

Our third method involves the linear majorization

$$\begin{aligned} f(x) &\leq f(y) + df(y)(x - y) \quad (6.5) \\ &= g(x \mid y) \end{aligned}$$

satisfied by any concave function $f(x)$. Once again we can replace the argument x by a vector-valued function $h(x)$.

Our fourth method applies to functions $f(x)$ with bounded curvature [6, 24]. Assuming that $f(x)$ is twice differentiable, we look for a matrix B satisfying $B \succeq d^2f(x)$ and $B \succ \mathbf{0}$ in the sense that $B - d^2f(x)$ is positive semidefinite for all x and B is positive definite. The quadratic bound principle then amounts to the majorization

$$\begin{aligned} f(x) &= f(y) + df(y)(x - y) \\ &\quad + (x-y)^* \int_0^1 d^2f[y + t(x-y)](1-t)\,dt\,(x-y) \\ &\leq f(y) + df(y)(x-y) + \frac{1}{2}(x-y)^*B(x-y) \quad (6.6) \\ &= g(x \mid y). \end{aligned}$$

Any of the preceding majorizations can be turned into minorizations by interchanging the adjectives convex and concave and positive definite and negative definite, respectively. Of course, there is an art to applying these methods just as there is an art to applying any mathematical principle. The methods hardly exhaust the possibilities for majorization and minorization. Problems 2, 8, 10, and 11 sketch other helpful techniques. Readers are also urged to consult the survey papers [24, 57, 70, 86] and the literature on the EM algorithm for a fuller discussion.

6.4 Allele Frequency Estimation

The ABO and Rh genetic loci are usually typed in matching blood donors to blood recipients. The ABO locus exhibits the three alleles A, B and O and the four observable phenotypes A, B, AB, and O. These phenotypes arise because each person inherits two alleles, one from his mother and one from his father, and the alleles A and B are genetically dominant to allele O. Dominance amounts to a masking of the O allele by the presence of an A or B allele. For instance, a person inheriting an A allele from one parent and an O allele from the other parent is said to have genotype A/O and is indistinguishable from a person inheriting an A allele from both parents. This second person has genotype A/A.

The MM algorithm for estimating the population frequencies or proportions of the three alleles involves an interplay between observed phenotypes and underlying unobserved genotypes. As just noted, both genotypes A/O and A/A correspond to the same phenotype A. Likewise, phenotype B corresponds to either genotype B/O or genotype B/B. Phenotypes AB and O correspond to the single genotypes A/B and O/O, respectively.

As a concrete example, Clarke et al. [19] noted that among their population sample of $n = 521$ duodenal ulcer patients, a total of $n_A = 186$ had phenotype A, $n_B = 38$ had phenotype B, $n_{AB} = 13$ had phenotype AB, and $n_O = 284$ had phenotype O. If we want to estimate the frequencies p_A, p_B, and p_O of the three different alleles from this sample, then we can employ the MM algorithm with the four phenotype counts as the observed data.

The likelihood of the data is given by the multinomial distribution in conjunction with the Hardy-Weinberg law of population genetics. This law specifies that each genotype frequency equals the product of the corresponding allele frequencies with an extra factor of 2 included to account for ambiguity in parental source when the two alleles differ. For example, genotype A/A has frequency p_A^2, and genotype A/O has frequency $2p_Ap_O$. These assumptions are summarized in the loglikelihood

$$\begin{aligned} f(p) &= n_A \ln\left(p_A^2 + 2p_Ap_O\right) + n_B \ln\left(p_B^2 + 2p_Bp_O\right) + n_{AB}\ln(2p_Ap_B) \\ &\quad + n_O \ln p_O^2 + \ln\binom{n}{n_A,\ n_B,\ n_{AB},\ n_O}. \end{aligned}$$

In maximum likelihood estimation we maximize this function of the allele frequencies subject to the equality constraint $p_A + p_B + p_O = 1$ and the nonnegativity constraints $p_A \geq 0$, $p_B \geq 0$, and $p_O \geq 0$.

The loglikelihood function $f(p)$ would be easy to maximize if it were not for the terms $\ln\left(p_A^2 + 2p_Ap_O\right)$ and $\ln\left(p_B^2 + 2p_Bp_O\right)$. In the MM algorithm we attack these functions using the convexity of the function $-\ln x$ and the majorization (6.3). This yields the minorization

$$\begin{aligned}
\ln\left(p_A^2 + 2p_A p_O\right) \;\geq\; & \frac{p_{mA}^2}{p_{mA}^2 + 2p_{mA}p_{mO}} \ln\left(\frac{p_{mA}^2 + 2p_{mA}p_{mO}}{p_{mA}^2} p_A^2\right) \\
& + \frac{2p_{mA}p_{mO}}{p_{mA}^2 + 2p_{mA}p_{mO}} \ln\left(\frac{p_{mA}^2 + 2p_{mA}p_{mO}}{2p_{mA}p_{mO}} 2p_A p_O\right).
\end{aligned}$$

A similar minorization applies to $\ln\left(p_B^2 + 2p_B p_O\right)$. These maneuvers have the virtue of separating parameters once we observe that logarithms turn products into sums.

Notationally, things become clearer if we introduce the abbreviations

$$\begin{aligned}
n_{mA/A} &= n_A \frac{p_{mA}^2}{p_{mA}^2 + 2p_{mA}p_{mO}} \\
n_{mA/O} &= n_A \frac{2p_{mA}p_{mO}}{p_{mA}^2 + 2p_{mA}p_{mO}}
\end{aligned}$$

and likewise for $n_{mB/B}$ and $n_{mB/O}$. We are now faced with maximizing the surrogate function

$$\begin{aligned}
g(p \mid p_{(m)}) &= n_{mA/A} \ln p_A^2 + n_{mA/O} \ln(2p_A p_O) + n_{mB/B} \ln p_B^2 \\
&\quad + n_{mB/O} \ln(2p_B p_O) + n_{AB} \ln(2p_A p_B) + n_O \ln p_O^2 + c,
\end{aligned}$$

where c is an irrelevant constant that depends on the previous iterate $p_{(m)}$ but not on the potential value p of the next iterate. This completes the minorization step of the algorithm.

The maximization step can be accomplished by introducing a Lagrange multiplier and finding a stationary point of the Lagrangian

$$\mathcal{L}(p, \lambda) = g(p \mid p_{(m)}) + \lambda(p_A + p_B + p_O - 1).$$

Here we ignore the nonnegativity constraints with the sure knowledge that they are inactive at the solution. Setting the partial derivatives of $\mathcal{L}(p, \lambda)$,

$$\begin{aligned}
\frac{\partial}{\partial p_A}\mathcal{L}(p, \lambda) &= \frac{2n_{mA/A}}{p_A} + \frac{n_{mA/O}}{p_A} + \frac{n_{AB}}{p_A} + \lambda \\
\frac{\partial}{\partial p_B}\mathcal{L}(p, \lambda) &= \frac{2n_{mB/B}}{p_B} + \frac{n_{mB/O}}{p_B} + \frac{n_{AB}}{p_B} + \lambda \\
\frac{\partial}{\partial p_O}\mathcal{L}(p, \lambda) &= \frac{n_{mA/O}}{p_O} + \frac{n_{mB/O}}{p_O} + \frac{2n_O}{p_O} + \lambda \\
\frac{\partial}{\partial \lambda}\mathcal{L}(p, \lambda) &= p_A + p_B + p_O - 1,
\end{aligned}$$

equal to 0 provides the unique stationary point of $\mathcal{L}(p, \lambda)$. The solution of the resulting equations is

$$p_{m+1,A} = \frac{2n_{mA/A} + n_{mA/O} + n_{AB}}{2n}$$

$$p_{m+1,B} = \frac{2n_{mB/B} + n_{mB/O} + n_{AB}}{2n}$$
$$p_{m+1,O} = \frac{n_{mA/O} + n_{mB/O} + 2n_O}{2n}.$$

In other words, the MM update is identical to a form of gene counting in which the unknown genotype counts are imputed based on the current allele frequency estimates [111]. In these updates, the denominator $2n$ is the total number of genes; the numerators are the current best guesses of the number of alleles of each type contained in the hidden and manifest genotypes.

Table 6.1 shows the progress of the MM iterates starting from the initial estimates $p_{0A} = 0.3$, $p_{0B} = 0.2$, and $p_{0O} = 0.5$. The MM updates are simple enough to carry out on a pocket calculator. Convergence occurs quickly in this example.

TABLE 6.1. Iterations for ABO Duodenal Ulcer Data

Iteration m	p_{mA}	p_{mB}	p_{mO}
0	0.3000	0.2000	0.5000
1	0.2321	0.0550	0.7129
2	0.2160	0.0503	0.7337
3	0.2139	0.0502	0.7359
4	0.2136	0.0501	0.7363
5	0.2136	0.0501	0.7363

6.5 Linear Regression

Because t^2 is a convex function, we can majorize each summand of the least squares criterion $\sum_{i=1}^n (y_i - x_i^*\theta)^2$ using inequality (6.4). It follows that

$$\sum_{i=1}^n (y_i - x_i^*\theta)^2 \le \sum_{i=1}^n \sum_j \alpha_{ij} \left[y_i - \frac{x_{ij}}{\alpha_{ij}}(\theta_j - \theta_{mj}) - x_i^*\theta_{(m)} \right]^2 = g(\theta \mid \theta_{(m)}),$$

with equality when $\theta = \theta_{(m)}$. Minimization of $g(\theta \mid \theta_{(m)})$ yields the updates

$$\theta_{m+1,j} = \theta_{mj} + \frac{\sum_{i=1}^n x_{ij}(y_i - x_i^*\theta_{(m)})}{\sum_{i=1}^n \frac{x_{ij}^2}{\alpha_{ij}}}$$

and avoids matrix inversion [86]. Although it seems intuitively reasonable to take $p = 1$ in choosing

$$\alpha_{ij} = \frac{|x_{ij}|^p}{(\sum_k |x_{ik}|^p)},$$

conceivably other values of p might perform better. In fact, it might accelerate convergence to alternate different values of p as the iterations proceed. For problems involving just a few parameters, this iterative scheme is clearly inferior to the usual single-step solution via matrix inversion. For high-dimensional problems, the iterative method becomes competitive.

Least squares estimation suffers from the fact that it is strongly influenced by observations far removed from their predicted values. In least absolute deviation regression, we replace $\sum_{i=1}^{n}(y_i - x_i^*\theta)^2$ by

$$h(\theta) = \sum_{i=1}^{n} \left| y_i - x_i^*\theta \right| = \sum_{i=1}^{n} |r_i(\theta)|, \tag{6.7}$$

where $r_i(\theta) = y_i - x_i^*\theta$ is the ith residual.We are now faced with minimizing a nondifferentiable function. Fortunately, the MM algorithm can be implemented by exploiting the concavity of the function $\sqrt{u}$ in inequality (6.5). Because

$$\sqrt{u} \le \sqrt{u_m} + \frac{u - u_m}{2\sqrt{u_m}},$$

we find that

$$\begin{aligned} h(\theta) &= \sum_{i=1}^{n} \sqrt{r_i^2(\theta)} \\ &\le h(\theta_{(m)}) + \frac{1}{2}\sum_{i=1}^{n} \frac{r_i^2(\theta) - r_i^2(\theta_{(m)})}{\sqrt{r_i^2(\theta_{(m)})}} \\ &= g(\theta \mid \theta_{(m)}). \end{aligned}$$

Minimizing $g(\theta \mid \theta_{(m)})$ is accomplished by minimizing the weighted sum of squares

$$\sum_{i=1}^{n} w_i(\theta_{(m)}) r_i(\theta)^2$$

with ith weight $w_i(\theta_{(m)}) = |r_i(\theta_{(m)})|^{-1}$. A slight variation of the usual argument for minimizing a sum of squares leads to the update

$$\theta_{(m+1)} = [X^*W(\theta_{(m)})X]^{-1}W(\theta_{(m)})X^*y,$$

where $W(\theta_{(m)})$ is the diagonal matrix with ith diagonal entry $w_i(\theta_{(m)})$. Unfortunately, the possibility that some weight $w_i(\theta_{(m)}) = \infty$ cannot be ruled out. Problem 9 suggests a simple remedy.

6.6 Bradley-Terry Model of Ranking

In the sports version of the Bradley and Terry model [11, 73], each team i in a league of teams is assigned a rank parameter $r_i > 0$. Assuming ties are impossible, team i beats team j with probability $r_i/(r_i + r_j)$. If this outcome occurs y_{ij} times during a season of play, then the probability of the whole season is

$$L(r) = \prod_{i,j} \left(\frac{r_i}{r_i + r_j}\right)^{y_{ij}},$$

assuming the games are independent. To rank the teams, we find the values $\hat{r}_i$ that maximize $f(r) = \ln L(r)$. The team with largest $\hat{r}_i$ is considered best, the team with smallest $\hat{r}_i$ is considered worst, and so forth. In view of the fact that $\ln u$ is concave, inequality (6.5) implies

$$\begin{aligned} f(r) &= \sum_{i,j} y_{ij}\Big[\ln r_i - \ln(r_i + r_j)\Big] \\ &\geq \sum_{i,j} y_{ij}\Big[\ln r_i - \ln(r_{mi} + r_{mj}) - \frac{r_i + r_j - r_{mi} - r_{mj}}{r_{mi} + r_{mj}}\Big] \\ &= g(r \mid r_m) \end{aligned}$$

with equality when $r = r_{(m)}$. Differentiating $g(r \mid r_{(m)})$ with respect to the ith component r_i of r and setting the result equal to 0 produces the next iterate

$$r_{m+1,i} = \frac{\sum_{j \neq i} y_{ij}}{\sum_{j \neq i}(y_{ij} + y_{ji})/(r_{mi} + r_{mj})}.$$

Because $L(r) = L(\beta r)$ for any $\beta > 0$, we constrain $r_1 = 1$ and omit the update $r_{m+1,1}$. In this example, the MM algorithm separates parameters and allows us to maximize $g(r \mid r_{(m)})$ parameter by parameter.

6.7 Linear Logistic Regression

In linear logistic regression, we observe a sequence of independent Bernoulli trials, each resulting in success or failure. The success probability of the ith trial

$$\pi_i(\theta) = \frac{e^{x_i^* \theta}}{1 + e^{x_i^* \theta}}$$

depends on a covariate vector x_i and parameter vector θ by analogy with linear regression. The observation y_i at trial i equals 1 for a success and 0 for a failure. In this notation, the likelihood of the data is

$$L(\theta) = \prod_i \pi_i(\theta)^{y_i}[1-\pi_i(\theta)]^{1-y_i}.$$

As usual in maximum likelihood estimation, we pass to the loglikelihood

$$f(\theta) = \sum_i [y_i \ln \pi_i(\theta) + (1-y_i)\ln[1-\pi_i(\theta)].$$

Straightforward calculations show

$$\begin{aligned} df(\theta) &= \sum_i [y_i - \pi_i(\theta)]x_i^* \\ d^2 f(\theta) &= -\sum_i \pi_i(\theta)[1-\pi_i(\theta)]x_i x_i^*. \end{aligned}$$

The loglikelihood $f(\theta)$ is therefore concave, and we seek to minorize it by a quadratic rather than majorize it by a quadratic as suggested in inequality (6.6). Hence, we must identify a matrix B such that B is negative definite and $B - d^2 f(x)$ is negative semidefinite for all x. In view of the scalar inequality $\pi(1-\pi) \leq \frac{1}{4}$, we take $B = -\frac{1}{4}\sum_i x_i x_i^*$. Maximization of the minorizing quadratic

$$f(\theta_{(m)}) + df(\theta_{(m)})(\theta - \theta_{(m)}) + \frac{1}{2}(\theta - \theta_{(m)})^* B(\theta - \theta_{(m)})$$

is a problem we have met before. It does involve inversion of the matrix B, but once we have computed B^{-1}, we can reuse it at every iteration.

6.8 Poisson Processes

In preparation for our exposition of transmission tomography in the next section, we now briefly review the theory of Poisson processes, a topic from probability of considerable interest in its own right. A Poisson process involves points randomly scattered in a region S of q-dimensional space R^q [53, 64, 72, 76]. The notion that the points are concentrated on average more in some regions than in others is captured by postulating an intensity function $\lambda(x) \geq 0$ on S. The expected number of points in a subregion T is given by the integral $\omega = \int_T \lambda(x)dx$. If $\omega = \infty$, then an infinite number of random points occur in T. If $\omega < \infty$, then a finite number of random points occur in T, and the probability that this number equals k is given by the Poisson probability

$$p_k(\omega) = \frac{\omega^k}{k!}e^{-\omega}.$$

Derivation of this formula depends critically on the assumption that the numbers N_{T_i} of random points in disjoint regions T_i are independent random variables. This basically means that knowing the values of some of the N_{T_i} tells one nothing about the values of the remaining N_{T_i}. The model also presupposes that random points never coincide.

The Poisson distribution has a peculiar relationship to the multinomial distribution. Suppose a Poisson random variable Z with mean ω represents the number of outcomes from some experiment, say an experiment involving a Poisson process. Let each outcome be independently classified in one of l categories, the kth of which occurs with probability p_k. Then the number of outcomes Z_k falling in category k is Poisson distributed with mean $\omega_k = p_k\omega$. Furthermore, the random variables $Z_1, \ldots, Z_l$ are independent. Conversely, if $Z = \sum_{k=1}^{l} Z_k$ is a sum of independent Poisson random variables Z_k with means $\omega_k = p_k\omega$, then conditional on $Z = n$, the vector $(Z_1, \ldots, Z_l)^*$ follows a multinomial distribution with n trials and cell probabilities $p_1, \ldots, p_l$. To prove the first two of these assertions, let $n = n_1 + \cdots + n_l$. Then

$$\begin{aligned}
\Pr(Z_1 = n_1, \ldots, Z_l = n_l) &= \frac{\omega^n}{n!}e^{-\omega}\binom{n}{n_1, \ldots, n_l}\prod_{k=1}^{l} p_k^{n_k} \\
&= \prod_{k=1}^{l} \frac{\omega_k^{n_k}}{n_k!}e^{-\omega_k} \\
&= \prod_{k=1}^{l} \Pr(Z_k = n_k).
\end{aligned}$$

To prove the converse, divide the last string of equalities by the probability $\Pr(Z = n) = \omega^n e^{-\omega}/n!$.

The random process of assigning points to categories is termed coloring in the stochastic process literature. When there are just two colors, and only random points of one of the colors are tracked, then the process is termed random thinning. We will see examples of both coloring and thinning in the next section.

6.9 Transmission Tomography

Problems in medical imaging often involve thousands of parameters. As an illustration of the MM algorithm, we treat maximum likelihood estimation in transmission tomography. Traditionally, transmission tomography images have been reconstructed by the methods of Fourier analysis. Fourier methods are fast but do not take into account the uncertainties of photon counts. Statistically based methods give better reconstructions with less patient exposure to harmful radiation.

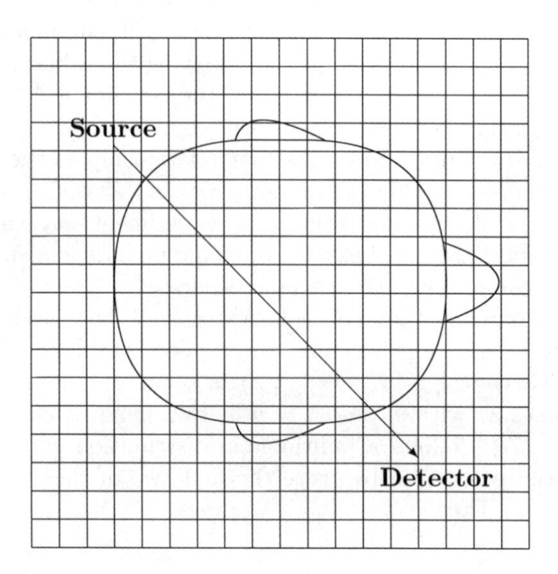

FIGURE 6.2. Cartoon of Transmission Tomography

The purpose of transmission tomography is to reconstruct the local attenuation properties of the object being imaged [59]. Attenuation is to be roughly equated with density. In medical applications, material such as bone is dense and stops or deflects X-rays (high-energy photons) better than soft tissue. With enough photons, even small gradations in soft tissue can be detected. A two-dimensional image is constructed from a sequence of photon counts. Each count corresponds to a projection line L drawn from an X-ray source through the imaged object to an X-ray detector. The average number of photons sent from the source along L to the detector is known in advance. The random number of photons actually detected is determined by the probability of a single photon escaping deflection or capture along L. Figure 6.2 shows one such projection line beamed through a cartoon of the human head.

To calculate this probability, let $\mu(x_1, x_2)$ be the intensity (or attenuation coefficient) of photon deflection or capture per unit length at position (x_1, x_2) in the plane. We can imagine that deflection or capture events occur completely randomly along L according to a Poisson process. The first such event effectively prevents the photon from being detected. Thus, the photon is detected with the Poisson probability $p_0(\omega) = e^{-\omega}$ of no such events, where

$$\omega = \int_L \mu(x_1, x_2) ds$$

is the line integral of $\mu(x_1, x_2)$ along L. In actual practice, X-rays are beamed through the object along a large number of different projection lines. We therefore face the inverse problem of reconstructing a function $\mu(x_1, x_2)$ in the plane from a large number of its measured line integrals. Imposing enough smoothness on $\mu(x_1, x_2)$, one can solve this classical deterministic problem by applying Radon transform techniques from Fourier analysis [59].

An alternative to the Fourier method is to pose an explicitly stochastic model and estimate its parameters by maximum likelihood [84, 85]. The MM algorithm suggests itself in this context. The stochastic model depends on dividing the object of interest into small nonoverlapping regions of constant attenuation called pixels. Typically the pixels are squares on a regular grid as depicted in Figure 6.2. The attenuation attributed to pixel j constitutes parameter θ_j of the model. Since there may be thousands of pixels, implementation of maximum likelihood algorithms such as scoring or Newton's method as discussed in the next chapter is out of the question.

To summarize our discussion, each observation Y_i is generated by beaming a stream of X-rays or high-energy photons from an X-ray source toward some detector on the opposite side of the object. The observation (or projection) Y_i counts the number of photons detected along the ith line of flight. Naturally, only a fraction of the photons are successfully transmitted from source to detector. If l_{ij} is the length of the segment of projection line i intersecting pixel j, then the probability of a photon escaping attenuation along projection line i is the exponentiated line integral $\exp(-\sum_j l_{ij}\theta_j)$.

In the absence of the intervening object, the number of photons generated and ultimately detected follows a Poisson distribution. We assume that the mean d_i of this distribution for projection line i is known. Ideally, detectors are long tubes aimed at the source. If a photon is deflected, then it is detected neither by the tube toward which it is initially headed nor by any other tube. In practice, many different detectors collect photons simultaneously from a single source. If we imagine coloring the tubes, then each photon is colored by the tube toward which it is directed. Each stream of colored photons is then thinned by capture or deflection. These considerations imply that the counts Y_i are independent and Poisson distributed with means $d_i \exp(-\sum_j l_{ij}\theta_j)$. It follows that we can express the loglikelihood of the observed data $Y_i = y_i$ as the finite sum

$$\sum_i \Big[-d_i e^{-\sum_j l_{ij}\theta_j} - y_i \sum_j l_{ij}\theta_j + y_i \ln d_i - \ln y_i! \Big]. \tag{6.8}$$

Omitting irrelevant constants, we can rewrite the loglikelihood (6.8) more succinctly as

$$L(\theta) \quad = \quad -\sum_i f_i(l_i^* \theta),$$

where $f_i(t)$ is the convex function $d_i e^{-t} + y_i t$ and $l_i^* \theta = \sum_j l_{ij}\theta_j$ is the inner product of the attenuation parameter vector θ and the vector of intersection lengths l_i for projection i.

To generate a surrogate function, we majorize each $f_i(l_i^*\theta)$ according to the recipe (6.3). This gives the surrogate function

$$g(\theta \mid \theta_{(m)}) \quad = \quad -\sum_i \sum_j \frac{l_{ij}\theta_{mj}}{l_i^*\theta_{(m)}} f_i\Big(\frac{l_i^*\theta_{(m)}}{\theta_{mj}}\theta_j\Big) \tag{6.9}$$

minorizing $L(\theta)$. By construction, maximization of $g(\theta \mid \theta_{(m)})$ separates into a sequence of one-dimensional problems, each of which can be solved approximately by one step of Newton's method. We will take up the details of this in the next chapter.

The images produced by maximum likelihood estimation in transmission tomography look grainy. The cure is to enforce image smoothness by penalizing large differences between estimated attenuation parameters of neighboring pixels. Geman and McClure [47] recommend multiplying the likelihood of the data by a Gibbs prior $\pi(\theta)$. Equivalently we add the log prior

$$\ln \pi(\theta) \quad = \quad -\gamma \sum_{\{j,k\} \epsilon N} w_{jk}\psi(\theta_j - \theta_k)$$

to the loglikelihood, where γ and the weights w_{jk} are positive constants, N is a set of unordered pairs $\{j,k\}$ defining a neighborhood system, and $\psi(r)$ is called a potential function. This function should be large whenever $|r|$ is large. Neighborhoods have limited extent. For instance, if the pixels are squares, we might define the weights by $w_{jk} = 1$ for orthogonal nearest neighbors sharing a side and $w_{jk} = 1/\sqrt{2}$ for diagonal nearest neighbors sharing only a corner. The constant γ scales the overall strength assigned to the prior. The sum $L(\theta) + \ln \pi(\theta)$ is called the logposterior function; its maximum is the posterior mode.

Choice of the potential function $\psi(r)$ is the most crucial feature of the Gibbs prior. It is convenient to assume that $\psi(r)$ is even and strictly convex. Strict convexity leads to the strict concavity of the log posterior function $L(\theta) + \ln \pi(\theta)$ and permits simple modification of the MM algorithm based on the surrogate function $g(\theta \mid \theta^n)$ defined by equation (6.9). Many potential functions exist satisfying these conditions. One natural example is $\psi(r) = r^2$. This choice unfortunately tends to deter the formation of boundaries. The gentler alternatives $\psi(r) = \sqrt{r^2 + \epsilon}$ for a small positive ϵ and $\psi(r) = \ln[\cosh(r)]$ are preferred in practice [52]. Problem 14 asks the reader to verify some of the properties of these two potential functions.

One adverse consequence of introducing a prior is that it couples pairs of parameters in the maximization step of the MM algorithm for finding the posterior mode. One can decouple the parameters by exploiting the convexity and evenness of the potential function $\psi(r)$ through the inequality

$$\begin{aligned}\psi(\theta_j - \theta_k) &= \psi\Big(\frac{1}{2}\Big[2\theta_j - \theta_{mj} - \theta_{mk}\Big] + \frac{1}{2}\Big[-2\theta_k + \theta_{mj} + \theta_{mk}\Big]\Big) \\ &\leq \frac{1}{2}\psi(2\theta_j - \theta_{mj} - \theta_{mk}) + \frac{1}{2}\psi(2\theta_k - \theta_{mj} - \theta_{mk}),\end{aligned}$$

which is strict unless $\theta_j + \theta_k = \theta_{mj} + \theta_{mk}$. This inequality allows us to redefine the surrogate function as

$$\begin{aligned}& g(\theta \mid \theta_{(m)}) \\ =& -\sum_i \sum_j \frac{l_{ij}\theta_{mj}}{l_i^* \theta_{(m)}} f_i\Big(\frac{l_i^*\theta_{(m)}}{\theta_{mj}}\theta_j\Big) \\ & -\frac{\gamma}{2} \sum_{\{j,k\} \epsilon N} w_{jk}[\psi(2\theta_j - \theta_{mj} - \theta_{mk}) + \psi(2\theta_k - \theta_{mj} - \theta_{mk})].\end{aligned}$$

Once again the parameters are separated, and the maximization step reduces to a sequence of one-dimensional problems. Maximizing $g(\theta \mid \theta_{(m)})$ drives the logposterior uphill and eventually leads to the posterior mode.

6.10 Problems

1. Prove that the majorization relation between functions is closed under the formation of sums, nonnegative products, limits, and composition with an increasing function.

2. Demonstrate the majorizing and minorizing inequalities

$$\begin{aligned}x^q &\leq q x_m^{q-1} x + (1-q) x_m^q \\ \ln x &\leq \frac{x}{x_m} + \ln x_m - 1 \\ \|x\| &\geq \frac{x_{(m)}^* x}{\|x_{(m)}\|} \\ 2xy &\leq \frac{x^2 y_m}{x_m} + \frac{y^2 x_m}{y_m} \\ \frac{1}{x} &\leq \frac{1}{x_m} - \frac{x - x_m}{x_m^2} + \frac{(x - x_m)^2}{c^3}.\end{aligned}$$

   Determine the relevant domains of each variable q, x, x_m, y, y_m, and c, and check that equality occurs in each of the inequalities when $x = x_m$ and $y = y_m$ [57].

3. Suppose $p \in [1, 2]$ and $x_m \neq 0$. Verify the majorizing inequality

$$|x|^p \leq \frac{p}{2}|x_m|^{p-2}x^2 + \Big(1 - \frac{p}{2}\Big)|x_m|^p.$$

This is important in least L_p regression.

4. Suppose the phenotypic counts in the ABO allele frequency estimation example satisfy $n_A + n_{AB} > 0$, $n_B + n_{AB} > 0$, and $n_O > 0$. Show that the loglikelihood is strictly concave and possesses a single global maximum on the interior of the feasible region.

5. In a genetic linkage experiment, 197 animals are randomly assigned to four categories according to the multinomial distribution with cell probabilities $\pi_1 = \frac{1}{2} + \frac{\theta}{4}$, $\pi_2 = \frac{1-\theta}{4}$, $\pi_3 = \frac{1-\theta}{4}$ and $\pi_4 = \frac{\theta}{4}$. If the corresponding observations are

$$\begin{aligned} y &= (y_1, y_2, y_3, y_4)^* \\ &= (125, 18, 20, 34)^*, \end{aligned}$$

then devise an MM algorithm and use it to estimate $\hat{\theta} = .6268$ [107]. (Hint: Minorize the part of the loglikelihood corresponding to the first category.)

6. Show that the loglikelihood $f(r) = \ln L(r)$ in Example 6.6 is concave under the reparameterization $r_i = e^{\theta_i}$.

7. A number μ is said to be a q quantile of the n numbers $x_1, \ldots, x_n$ if it satisfies

$$\begin{aligned} \frac{1}{n} \sum_{i: x_i \le \mu} 1 &\ge q \\ \frac{1}{n} \sum_{i: x_i \ge \mu} 1 &\ge 1 - q. \end{aligned}$$

If we define

$$\rho_q(r) = \begin{cases} qr & r \ge 0 \\ (1-q)|r| & r < 0, \end{cases}$$

then demonstrate that μ is a q quantile if and only if μ minimizes the function $f_q(\mu) = \sum_{i=1}^n \rho_q(x_i - \mu)$. Medians correspond to the case $q = 1/2$.

8. Continuing Problem 7, show that the function $\rho_q(r)$ is majorized by the quadratic

$$\zeta_q(r \mid r_m) = \frac{1}{4} \left[\frac{r^2}{|r_m|} + (4q - 2)r + |r_m| \right].$$

Deduce from this majorization the MM algorithm

$$\begin{aligned} \mu_{m+1} &= \frac{n(2q-1) + \sum_{i=1}^n w_{mi} x_i}{\sum_{i=1}^n w_{mi}} \\ w_{mi} &= \frac{1}{|x_i - \mu_m|} \end{aligned}$$

for finding a q quantile. This interesting algorithm involves no sorting, only arithmetic operations.

9. Suppose we minimize the function

$$h_\epsilon(\theta) = \sum_{i=1}^{p} \left\{ \left[y_i - \sum_{j=1}^{q} x_{ij}\theta_j \right]^2 + \epsilon \right\}^{1/2}$$

instead of the function $h(\theta)$ in equation (6.7) for a small, positive number ϵ. Show that the same MM algorithm applies with revised weights $w_i(\theta_{(m)}) = 1/\sqrt{r_i(\theta_{(m)})^2 + \epsilon}$.

10. In the Bradley-Terry model of Section 6.6, suppose we want to include the possibility of ties. One way of doing this is to write the probabilities of the three outcomes of i versus j as

$$\begin{aligned}
\Pr(\text{i wins}) &= \frac{r_i}{r_i + r_j + \theta\sqrt{r_i r_j}} \\
\Pr(\text{i ties}) &= \frac{\theta\sqrt{r_i r_j}}{r_i + r_j + \theta\sqrt{r_i r_j}} \\
\Pr(\text{i loses}) &= \frac{r_j}{r_i + r_j + \theta\sqrt{r_i r_j}},
\end{aligned}$$

where $\theta > 0$ is an additional parameter to be estimated. Let y_{ij} represent the number of times i beats j and t_{ij} the number of times i ties j. Prove that the loglikelihood of the data is

$$\begin{aligned}
&L(\theta, r) \\
&= \frac{1}{2}\sum_{i,j}\left(2y_{ij}\ln\frac{r_i}{r_i + r_j + \theta\sqrt{r_i r_j}} + t_{ij}\ln\frac{\theta\sqrt{r_i r_j}}{r_i + r_j + \theta\sqrt{r_i r_j}}\right).
\end{aligned}$$

One way of maximizing $L(\theta, r)$ is to alternate between updating θ and r. Both of these updates can be derived from the perspective of the MM algorithm. Two minorizations are now involved. The first proceeds using the convexity of $-\ln t$ just as in the text. This produces a function involving $-\sqrt{r_i r_j}$ terms. Demonstrate the minorization

$$-\sqrt{r_i r_j} \geq -\frac{r_i}{2}\sqrt{\frac{r_{mj}}{r_{mi}}} - \frac{r_j}{2}\sqrt{\frac{r_{mi}}{r_{mj}}},$$

and use it to minorize $L(\theta, r)$. Finally, determine $r_{(m+1)}$ for θ fixed at θ_m and θ_{m+1} for r fixed at $r_{(m+1)}$. The details are messy, but the overall strategy is straightforward.

11. In the linear logistic model of Section 6.7, it is possible to separate parameters and avoid matrix inversion altogether. In constructing a minorizing function, first prove the inequality

$$\begin{aligned} \ln[1-\pi(\theta)] &= -\ln\left(1+e^{x_i^*\theta}\right) \\ &\geq -\ln\left(1+e^{x_i^*\theta_{(m)}}\right) - \frac{e^{x_i^*\theta}-e^{x_i^*\theta_{(m)}}}{1+e^{x_i^*\theta_{(m)}}}, \end{aligned}$$

with equality when $\theta = \theta_{(m)}$. This eliminates the log terms. Now apply the arithmetic-geometric mean inequality to the exponential functions $e^{x_i^*\theta}$ to separate parameters. Assuming that θ has n components and that there are k observations, show that these maneuvers lead to the minorizing function

$$g(\theta \mid \theta_{(m)}) = -\frac{1}{n}\sum_{i=1}^{k}\frac{e^{x_i^*\theta_{(m)}}}{1+e^{x_i^*\theta_{(m)}}}\sum_{j=1}^{n}e^{nx_{ij}(\theta_j-\theta_{mj})} + \sum_{i=1}^{k} y_i x_i^*\theta$$

up to a constant that does not depend on θ. Finally, prove that maximizing $g(\theta \mid \theta_{(m)})$ consists in solving the transcendental equation

$$-\sum_{i=1}^{k}\frac{e^{x_i^*\theta_{(m)}}x_{ij}e^{-nx_{ij}\theta_{mj}}}{1+e^{x_i^*\theta_{(m)}}}e^{nx_{ij}\theta_j} + \sum_{i=1}^{k} y_i x_{ij} = 0$$

for each j. This can be accomplished numerically.

12. Show that the loglikelihood (6.8) for the transmission tomography model is concave. State a necessary condition for strict concavity in terms of the number of pixels and the number of projections.

13. In the maximization phase of the MM algorithm for transmission tomography without a smoothing prior, demonstrate that the exact solution of the one-dimensional equation

$$\frac{\partial}{\partial\theta_j} g(\theta \mid \theta_{(m)}) = 0$$

exists and is positive when $\sum_i l_{ij}d_i > \sum_i l_{ij}y_i$. Why would this condition typically hold in practice?

14. Prove that the functions $\psi(r) = \sqrt{r^2+\epsilon}$ and $\psi(r) = \ln[\cosh(r)]$ are even, strictly convex, infinitely differentiable, and asymptotic to $|r|$ as $|r| \to \infty$.

7
The EM Algorithm

7.1 Introduction

Maximum likelihood is the dominant form of estimation in applied statistics. Because closed-form solutions to likelihood equations are the exception rather than the rule, numerical methods for finding maximum likelihood estimates are of paramount importance. In this chapter we study maximum likelihood estimation by the EM algorithm [25, 88, 93], a special case of the MM algorithm. At the heart of every EM algorithm is some notion of missing data. Data can be missing in the ordinary sense of a failure to record certain observations on certain cases. Data can also be missing in a theoretical sense. We can think of the E, or expectation, step of the algorithm as filling in the missing data. This action replaces the loglikelihood of the observed data by a minorizing function. This surrogate function is then maximized in the M step. Because the surrogate function is usually much simpler than the likelihood, we can often solve the M step analytically. The price we pay for this simplification is that the EM algorithm is iterative. Reconstructing the missing data is bound to be slightly wrong if the parameters do not already equal their maximum likelihood estimates.

One of the advantages of the EM algorithm is its numerical stability. As an MM algorithm, any EM algorithm leads to a steady increase in the likelihood of the observed data. Thus, the EM algorithm avoids wildly overshooting or undershooting the maximum of the likelihood along its current direction of search. Besides this desirable feature, the EM handles parameter constraints gracefully. Constraint satisfaction is by definition built into the solution of the M step. In contrast, competing methods of maximization must incorporate special techniques to cope with parameter constraints. The EM shares some of the negative features of the more general MM algorithm. For example, the EM algorithm often converges at an excruciatingly slow rate in a neighborhood of the maximum point. This rate directly reflects the amount of missing data in a problem. In the absence of concavity, there is also no guarantee that the EM algorithm will converge to the global maximum. The global maximum can usually be reached by starting the parameters at good but suboptimal estimates such as method-of-moments estimates or by choosing multiple random starting points.

7.2 Definition of the EM Algorithm

A sharp distinction is drawn in the EM algorithm between the observed, incomplete data Y and the unobserved, complete data X of a statistical experiment [25, 88, 115]. Some function $t(X) = Y$ collapses X onto Y. For instance, if we represent X as (Y, Z), with Z as the missing data, then t is simply projection onto the Y-component of X. It should be stressed that the missing data can consist of more than just observations missing in the ordinary sense. In fact, the definition of X is left up to the intuition and cleverness of the statistician. The general idea is to choose X so that maximum likelihood estimation becomes trivial for the complete data.

The complete data are assumed to have a probability density $f(X \mid \theta)$ that is a function of a parameter vector θ as well as of X. In the E step of the EM algorithm, we calculate the conditional expectation

$$Q(\theta \mid \theta_{(n)}) = \mathrm{E}[\ln f(X \mid \theta) \mid Y, \theta_{(n)}].$$

Here $\theta_{(n)}$ is the current estimated value of θ. In the M step, we maximize $Q(\theta \mid \theta_{(n)})$ with respect to θ. This yields the new parameter estimate $\theta_{(n+1)}$, and we repeat this two-step process until convergence occurs. Note that θ and $\theta_{(n)}$ play fundamentally different roles in $Q(\theta \mid \theta_{(n)})$.

If $\ln g(Y \mid \theta)$ denotes the loglikelihood of the observed data, then the EM algorithm enjoys the ascent property

$$\ln g(Y \mid \theta_{(n+1)}) \geq \ln g(Y \mid \theta_{(n)}).$$

Proof of this assertion unfortunately involves measure theory, so some readers may want to take it on faith and skip the rest of this section. A necessary preliminary is the following well-known inequality from statistics.

Proposition 7.2.1 (Information Inequality) *Let h and k be probability densities with respect to a measure μ. Suppose $h > 0$ and $k > 0$ almost* *everywhere relative to* μ. *If* E_h *denotes expectation with respect to the probability measure* $h d\mu$, *then* $\mathrm{E}_h(\ln h) \geq \mathrm{E}_h(\ln k)$, *with equality if and only if* $h = k$ *almost everywhere relative to* μ.

Proof: Because $-\ln(w)$ is a strictly convex function on $(0, \infty)$, Proposition 5.5.1 applied to the random variable k/h implies

$$\begin{aligned} \mathrm{E}_h(\ln h) - \mathrm{E}_h(\ln k) &= \mathrm{E}_h\Big(-\ln\frac{k}{h}\Big) \\ &\geq -\ln \mathrm{E}_h\Big(\frac{k}{h}\Big) \\ &= -\ln \int \frac{k}{h} h d\mu \\ &= -\ln \int k d\mu \\ &= 0. \end{aligned}$$

Equality holds if and only if $k/h = \mathrm{E}_h(k/h)$ almost everywhere relative to μ. This necessary and sufficient condition is equivalent to $h = k$ since $\mathrm{E}_h(k/h) = 1$. ■

To prove the ascent property of the EM algorithm, it suffices to demonstrate the minorization inequality

$$\ln g(y \mid \theta) \geq Q(\theta \mid \theta_{(n)}) + \ln g(y \mid \theta_{(n)}) - Q(\theta_{(n)} \mid \theta_{(n)}),$$

where $Q(\theta \mid \theta_{(n)}) = \mathrm{E}[\ln f(X \mid \theta) \mid Y = y, \theta_{(n)}]$. With this end in mind, note that both $f(x \mid \theta)/g(y \mid \theta)$ and $f(x \mid \theta_{(n)})/g(y \mid \theta_{(n)})$ are conditional densities of X on $\{x : t(x) = y\}$ with respect to some measure μ_y. The information inequality now indicates that

$$\begin{aligned} Q(\theta \mid \theta_{(n)}) - \ln g(y \mid \theta) &= \mathrm{E}\Big(\ln\Big[\frac{f(X \mid \theta)}{g(Y \mid \theta)}\Big] \mid Y = y, \theta_{(n)}\Big) \\ &\leq \mathrm{E}\Big(\ln\Big[\frac{f(X \mid \theta_{(n)})}{g(Y \mid \theta_{(n)})}\Big] \mid Y = y, \theta_{(n)}\Big) \\ &= Q(\theta_{(n)} \mid \theta_{(n)}) - \ln g(y \mid \theta_{(n)}). \end{aligned}$$

Duplicating the proof of the MM inequality (6.2), we now argue that

$$\begin{aligned} \ln g(y \mid \theta_{(n+1)}) &\geq Q(\theta_{(n+1)} \mid \theta_{(n)}) + \ln g(y \mid \theta_{(n+1)}) - Q(\theta_{(n+1)} \mid \theta_{(n)}) \\ &\geq Q(\theta_{(n)} \mid \theta_{(n)}) + \ln g(y \mid \theta_{(n)}) - Q(\theta_{(n)} \mid \theta_{(n)}) \\ &= \ln g(y \mid \theta_{(n)}). \end{aligned}$$

Strict inequality occurs in this ascent inequality when the conditional density $f(x \mid \theta)/g(y \mid \theta)$ differs at the parameter points $\theta_{(n)}$ and $\theta_{(n+1)}$ or

$$Q(\theta_{(n+1)} \mid \theta_{(n)}) > Q(\theta_{(n)} \mid \theta_{(n)}).$$

The preceding proof is a little vague as to the meaning of the conditional density $f(x \mid \theta)/g(y \mid \theta)$ and its associated measure μ_y. Commonly the complete data decomposes as $X = (Y, Z)$, where Z is considered the missing data and $t(Y, Z) = Y$ is projection onto the observed data. Suppose (Y, Z) has joint density $f(y, z \mid \theta)$ relative to a product measure $\omega \times \mu\,(y, z)$; ω and μ are typically Lebesgue measure or counting measure. In this framework, we define $g(y \mid \theta) = \int f(y, z, \theta) d\mu(z)$ and set $\mu_y = \mu$. The function $g(y \mid \theta)$ serves as a density relative to ω. To check that these definitions make sense, it suffices to prove that $\int h(y, z) f(y, z \mid \theta)/g(y \mid \theta) d\mu(z)$ is a version of the conditional expectation $\mathrm{E}[h(Y, Z) \mid Y = y]$ for every well-behaved function $h(y, z)$. This assertion can be verified by showing

$$\mathrm{E}\{1_S(Y)\, \mathrm{E}[h(Y, Z) \mid Y]\} = \mathrm{E}[1_S(Y) h(Y, Z)]$$

for every measurable set S. With

$$\mathrm{E}[h(Y, Z) \mid Y = y] = \int h(y, z) \frac{f(y, z \mid \theta)}{g(y \mid \theta)} d\mu(z),$$

we calculate

$$\begin{aligned}
\mathrm{E}\{1_S(Y)\, \mathrm{E}[h(Y, Z) \mid Y]\} &= \int_S \int h(y, z) \frac{f(y, z \mid \theta)}{g(y \mid \theta)} d\mu(z) g(y \mid \theta) d\omega(y) \\
&= \int_S \int h(y, z) f(y, z \mid \theta) d\mu(z) d\omega(y) \\
&= \mathrm{E}[1_S(Y) h(Y, Z)].
\end{aligned}$$

Thus in this situation, $f(x \mid \theta)/g(y \mid \theta)$ is indeed the conditional density of X given $Y = y$.

7.3 Allele Frequency Estimation

It is instructive to compare the EM and MM algorithms on identical problems. Even when the two algorithms specify the same iteration scheme, the differences in deriving the algorithms are illuminating. Consider the ABO allele frequency estimation problem of Section 6.4. From the EM perspective, the complete data are genotype counts rather than phenotype counts. In passing from the complete data to the observed data, nature collapses genotypes A/A and A/O into phenotype A and genotypes B/B and B/O into phenotype B. In view of the Hardy-Weinberg equilibrium law, the complete data multinomial loglikelihood becomes

$$\begin{aligned}
\ln f(X \mid p) &= n_{A/A} \ln p_A^2 + n_{A/O} \ln(2p_A p_O) + n_{B/B} \ln p_B^2 \\
&\quad + n_{B/O} \ln(2p_B p_O) + n_{AB} \ln(2p_A p_B) + n_O \ln p_O^2 \\
&\quad + \ln \binom{n}{n_{A/A},\, n_{A/O},\, n_{B/B},\, n_{B/O},\, n_{AB},\, n_O}. \qquad (7.1)
\end{aligned}$$

In the E step of the EM algorithm we take the expectation of $\ln f(X \mid p)$ conditional on the observed counts n_A, n_B, n_{AB}, and n_O and the current parameter vector $p_{(m)} = (p_{mA}, p_{mB}, p_{mO})^*$. It is obvious that

$$\begin{aligned} \mathrm{E}(n_{AB} \mid Y, p_{(m)}) &= n_{AB} \\ \mathrm{E}(n_O \mid Y, p_{(m)}) &= n_O. \end{aligned}$$

A moment's reflection also yields

$$\begin{aligned} n_{mA/A} &= \mathrm{E}(n_{A/A} \mid Y, p_{(m)}) \\ &= n_A \frac{p_{mA}^2}{p_{mA}^2 + 2p_{mA}p_{mO}} \\ n_{mA/O} &= \mathrm{E}(n_{A/O} \mid Y, p_{(m)}) \\ &= n_A \frac{2p_{mA}p_{mO}}{p_{mA}^2 + 2p_{mA}p_{mO}}. \end{aligned}$$

The conditional expectations $n_{mB/B}$ and $n_{mB/O}$ are given by similar expressions. The corresponding $Q(p \mid p_{(m)})$ function derived from the complete data likelihood (7.1) substitutes $n_{mA/A}$ for $n_{A/A}$ and so forth. This action yields exactly the same surrogate function as the MM algorithm, and maximization of $Q(p \mid p_m)$ proceeds as described earlier. One of the advantages of the EM derivation is that it explicitly reveals the nature of the conditional expectations $n_{mA/A}$, $n_{mA/O}$, $n_{mB/B}$, and $n_{mB/O}$.

7.4 Transmission Tomography

Derivation of the EM algorithm for transmission tomography is more interesting. In this instance, the EM and MM algorithms differ. The MM algorithm is easier to derive and computationally more efficient. In other examples, the opposite is true.

In the transmission tomography example of Section 6.9, it is natural to view the missing data as the number of photons X_{ij} entering each pixel j along each projection line i. These random variables supplemented by the observations Y_i constitute the complete data. If projection line i does not intersect pixel j, then $X_{ij} = 0$. Although X_{ij} and $X_{ij'}$ are not independent, the collection $\{X_{ij}\}_j$ indexed by projection i is independent of the collection $\{X_{i'j}\}_j$ indexed by another projection i'. This allows us to work projection by projection in writing the complete data likelihood. We will therefore temporarily drop the projection subscript i and relabel pixels, starting with pixel 1 adjacent to the source and ending with pixel $m-1$ adjacent to the detector. In this notation X_1 is the number of photons leaving the source, X_j is the number of photons entering pixel j, and $X_m = Y$ is the number of photons detected.

By assumption X_1 follows a Poisson distribution with mean d. Conditional on $X_1, \ldots, X_j$, the random variable X_{j+1} is binomially distributed with X_j trials and success probability $e^{-l_j\theta_j}$. In other words, each of the X_j photons entering pixel j behaves independently and has a chance $e^{-l_j\theta_j}$ of avoiding attenuation in pixel j. It follows that the complete data loglikelihood for the current projection is

$$\begin{aligned} & -d + X_1 \ln d - \ln X_1! \\ & + \sum_{j=1}^{m-1} \left[\ln \binom{X_j}{X_{j+1}} + X_{j+1} \ln e^{-l_j\theta_j} + (X_j - X_{j+1}) \ln(1 - e^{-l_j\theta_j}) \right]. \end{aligned} \tag{7.2}$$

To perform the E step of the EM algorithm, we need only compute the conditional expectations $\mathrm{E}(X_j \mid X_m = y, \theta)$, $j = 1, \ldots, m$. The conditional expectations of other terms such as $\ln \binom{X_j}{X_{j+1}}$ appearing in (7.2) are irrelevant in the subsequent M step.

Reasoning as above, we infer that the unconditional mean of X_j is

$$\begin{aligned} \mu_j &= \mathrm{E}(X_j) \\ &= d e^{-\sum_{k=1}^{j-1} l_k\theta_k} \end{aligned}$$

and that the distribution of X_m conditional on X_j is binomial with X_j trials and success probability

$$\frac{\mu_m}{\mu_j} = e^{-\sum_{k=j}^{m-1} l_k\theta_k}.$$

Hence, the joint probability of X_j and X_m reduces to

$$\Pr(X_j = x_j, X_m = x_m) = e^{-\mu_j} \frac{\mu_j^{x_j}}{x_j!} \binom{x_j}{x_m} \left(\frac{\mu_m}{\mu_j}\right)^{x_m} \left(1 - \frac{\mu_m}{\mu_j}\right)^{x_j - x_m},$$

and the conditional probability of X_j given X_m becomes

$$\begin{aligned} \Pr(X_j = x_j \mid X_m = x_m) &= \frac{e^{-\mu_j} \frac{\mu_j^{x_j}}{x_j!} \binom{x_j}{x_m} \left(\frac{\mu_m}{\mu_j}\right)^{x_m} \left(1 - \frac{\mu_m}{\mu_j}\right)^{x_j - x_m}}{e^{-\mu_m} \frac{\mu_m^{x_m}}{x_m!}} \\ &= e^{-(\mu_j - \mu_m)} \frac{(\mu_j - \mu_m)^{x_j - x_m}}{(x_j - x_m)!}. \end{aligned}$$

In other words, conditional on X_m, the difference $X_j - X_m$ follows a Poisson distribution with mean $\mu_j - \mu_m$. This implies in particular that

$$\begin{aligned} \mathrm{E}(X_j \mid X_m) &= \mathrm{E}(X_j - X_m \mid X_m) + X_m \\ &= \mu_j - \mu_m + X_m. \end{aligned}$$

Reverting to our previous notation, it is now possible to assemble the function $Q(\theta \mid \theta_{(n)})$ of the E step. Define

$$\begin{aligned} M_{ij} &= d_i(e^{-\sum_{k \in S_{ij}} l_{ik}\theta_{nk}} - e^{-\sum_k l_{ik}\theta_{nk}}) + y_i \\ N_{ij} &= d_i(e^{-\sum_{k \in S_{ij} \cup \{j\}} l_{ik}\theta_{nk}} - e^{-\sum_k l_{ik}\theta_{nk}}) + y_i, \end{aligned}$$

where S_{ij} is the set of pixels between the source and pixel j along projection i. If j' is the next pixel after pixel j along projection i, then

$$\begin{aligned} M_{ij} &= \mathrm{E}(X_{ij} \mid Y_i = y_i, \theta_{(n)}) \\ N_{ij} &= \mathrm{E}(X_{ij'} \mid Y_i = y_i, \theta_{(n)}). \end{aligned}$$

In view of expression (7.2), we find

$$Q(\theta \mid \theta_{(n)}) = \sum_i \sum_j \Big[-N_{ij} l_{ij} \theta_j + (M_{ij} - N_{ij}) \ln(1 - e^{-l_{ij}\theta_j}) \Big]$$

up to an irrelevant constant.

If we try to maximize $Q(\theta \mid \theta_{(n)})$ by setting its partial derivatives equal to 0, we get for pixel j the equation

$$-\sum_i N_{ij} l_{ij} + \sum_i \frac{(M_{ij} - N_{ij}) l_{ij}}{e^{l_{ij}\theta_j} - 1} = 0. \tag{7.3}$$

This is an intractable transcendental equation in the single variable θ_j, and the M step must be solved numerically, say by Newton's method. It is straightforward to check that the left-hand side of equation (7.3) is strictly decreasing in θ_j and has exactly one positive solution. Thus, the EM algorithm like the MM algorithm has the advantages of decoupling the parameters in the likelihood equations and of satisfying the natural boundary constraints $\theta_j \geq 0$. The MM algorithm is preferable to the EM algorithm because the MM algorithm involves far fewer exponentiations in defining its surrogate function.

7.5 Factor Analysis

In some instances, the missing data framework of the EM algorithm offers the easiest way to exploit convexity in deriving an MM algorithm. The complete data for a given problem is often fairly natural, and the difficulty in deriving an EM algorithm shifts toward specifying the E step. Statisticians are particularly adept at calculating complicated conditional expectations connected with sampling distributions. We now illustrate these truths for estimation in factor analysis. Factor analysis explains the covariation among the components of a random vector by approximating the vector by a linear transformation of a small number of uncorrelated factors. Because factor analysis models usually involve normally distributed random vectors, we review some basic facts about the multivariate normal distribution in the Appendix.

For the sake of notational convenience, we now extend the expectation and variance operators to random vectors. The expectation of a random vector $X = (X_1, \ldots, X_n)^*$ is defined componentwise by

$$\mathrm{E}(X) = \begin{pmatrix} \mathrm{E}[X_1] \\ \vdots \\ \mathrm{E}[X_n] \end{pmatrix}.$$

Linearity carries over from the scalar case in the sense that

$$\begin{aligned} \mathrm{E}(X+Y) &= \mathrm{E}(X) + \mathrm{E}(Y) \\ \mathrm{E}(MX) &= M\,\mathrm{E}(X) \end{aligned}$$

for a compatible random vector Y and a compatible matrix M. The same componentwise conventions hold for the expectation of a random matrix and the variances and covariances of a random vector. Thus, we can express the variance-covariance matrix of a random vector X as

$$\mathrm{Var}(X) = \mathrm{E}\{[X - \mathrm{E}(X)][X - \mathrm{E}(X)]^*\} = \mathrm{E}(XX^*) - \mathrm{E}(X)\,\mathrm{E}(X)^*.$$

These notational choices produce many other compact formulas. For instance, the random quadratic form X^*MX has expectation

$$\mathrm{E}(X^*MX) = \mathrm{tr}[M\,\mathrm{Var}(X)] + \mathrm{E}(X)^*M\,\mathrm{E}(X). \tag{7.4}$$

To verify this assertion, observe that

$$\begin{aligned} \mathrm{E}(X^*MX) &= \mathrm{E}\Big(\sum_{ij} X_i m_{ij} X_j\Big) \\ &= \sum_{ij} m_{ij}\,\mathrm{E}(X_i X_j) \\ &= \sum_{ij} m_{ij}[\mathrm{Cov}(X_i, X_j) + \mathrm{E}(X_i)\,\mathrm{E}(X_j)] \\ &= \mathrm{tr}[M\,\mathrm{Var}(X)] + \mathrm{E}(X)^*M\,\mathrm{E}(X). \end{aligned}$$

The classical factor analysis model deals with l independent multivariate observations of the form

$$Y_{(k)} = \mu + FX_{(k)} + U_{(k)}.$$

Here the $p \times q$ factor loading matrix F transforms the unobserved factor score $X_{(k)}$ into the observed $Y_{(k)}$. The random vector $U_{(k)}$ represents random measurement error. Typically, q is much smaller than p. The random vectors $X_{(k)}$ and $U_{(k)}$ are independent and normally distributed with means and variances

$$\begin{aligned} \mathrm{E}(X_{(k)}) &= \mathbf{0}, & \mathrm{Var}(X_{(k)}) &= I \\ \mathrm{E}(U_{(k)}) &= \mathbf{0}, & \mathrm{Var}(U_{(k)}) &= D, \end{aligned}$$

where I is the $q \times q$ identity matrix and D is a $p \times p$ diagonal matrix with ith diagonal entry d_i. The entries of the mean vector μ, the factor loading matrix F, and the diagonal matrix D constitute the parameters of the model. For a particular realization $y_{(1)}, \ldots, y_{(l)}$ of the model, the maximum likelihood estimation of μ is simply the sample mean $\hat{\mu} = \bar{y}$. This fact is a consequence of the reasoning given in Example 5.4.5. Therefore, we replace each $y_{(k)}$ by $y_{(k)} - \bar{y}$ and focus on estimating F and D.

The random vector $(X^*_{(k)}, Y^*_{(k)})^*$ is the obvious choice of the complete data for case k. If $f(x_{(k)})$ is the density of $X_{(k)}$ and $g(y_{(k)} \mid x_{(k)})$ is the conditional density of $Y_{(k)}$ given $X_{(k)} = x_{(k)}$, then the complete data loglikelihood can be expressed as

$$\begin{aligned} & \sum_{k=1}^{l} \ln f(x_{(k)}) + \sum_{k=1}^{l} \ln g(y_{(k)} \mid x_{(k)}) \\ = & -\frac{l}{2} \ln \det I - \frac{1}{2} \sum_{k=1}^{l} x^*_{(k)} x_{(k)} - \frac{l}{2} \ln \det D \qquad (7.5) \\ & -\frac{1}{2} \sum_{k=1}^{l} (y_{(k)} - F x_{(k)})^* D^{-1} (y_{(k)} - F x_{(k)}). \end{aligned}$$

We can simplify this by noting that $\ln \det I = 0$ and $\ln \det D = \sum_{i=1}^{p} \ln d_i$.

The key to performing the E step is to note that $(X^*_{(k)}, Y^*_{(k)})^*$ follows a multivariate normal distribution with variance matrix

$$\mathrm{Var}\begin{pmatrix} X_{(k)} \\ Y_{(k)} \end{pmatrix} = \begin{pmatrix} I & F^* \\ F & FF^* + D \end{pmatrix}.$$

Equation (A.1) of the Appendix then permits us to calculate the conditional expectation

$$\begin{aligned} v_{(k)} &= \mathrm{E}(X_{(k)} \mid Y_{(k)} = y_{(k)}) \\ &= F^*(FF^* + D)^{-1} y_{(k)} \end{aligned}$$

and conditional variance

$$\begin{aligned} A_{(k)} &= \mathrm{Var}(X_{(k)} \mid Y_{(k)} = y_{(k)}) \\ &= I - F^*(FF^* + D)^{-1} F, \end{aligned}$$

given the observed data and the current values of the matrices F and D. Combining these results with equation (7.4) yields

$$\begin{aligned}
& \mathrm{E}[(Y_{(k)} - FX_{(k)})^* D^{-1}(Y_{(k)} - FX_{(k)}) \mid Y_{(k)} = y_{(k)}] \\
= \; & \mathrm{tr}(D^{-1} F A_{(k)} F^*) + (y_{(k)} - Fv_{(k)})^* D^{-1}(y_{(k)} - Fv_{(k)}) \\
= \; & \mathrm{tr}\{D^{-1}[F A_{(k)} F^* + (y_{(k)} - Fv_{(k)})(y_{(k)} - Fv_{(k)})^*]\}.
\end{aligned}$$

If we define

$$\begin{aligned}
\Lambda &= \sum_{k=1}^{l} [A_{(k)} + v_{(k)} v_{(k)}^*] \\
\Gamma &= \sum_{k=1}^{l} v_{(k)} y_{(k)}^* \\
\Omega &= \sum_{k=1}^{l} y_{(k)} y_{(k)}^*
\end{aligned}$$

and take conditional expectations in equation (7.5), then we can write the surrogate function of the E step as

$$\begin{aligned}
& Q(F, D \mid F_{(n)}, D_{(n)}) \\
= \; & -\frac{l}{2} \sum_{i=1}^{p} \ln d_i - \frac{1}{2} \mathrm{tr}[D^{-1}(F\Lambda F^* - F\Gamma - \Gamma^* F^* + \Omega)],
\end{aligned}$$

omitting an additive constant that does not depend on either F or D.

To perform the M step, we first maximize $Q(F, D \mid F_{(n)}, D_{(n)})$ with respect to F, holding D fixed. We can do so by completing the square in the trace

$$\begin{aligned}
& \mathrm{tr}[D^{-1}(F\Lambda F^* - F\Gamma - \Gamma^* F^* + \Omega)] \\
= \; & \mathrm{tr}[D^{-1}(F - \Gamma^*\Lambda^{-1})\Lambda(F - \Gamma^*\Lambda^{-1})^*] + \mathrm{tr}[D^{-1}(\Omega - \Gamma^*\Lambda^{-1}\Gamma)] \\
= \; & \mathrm{tr}[D^{-\frac{1}{2}}(F - \Gamma^*\Lambda^{-1})\Lambda(F - \Gamma^*\Lambda^{-1})^* D^{-\frac{1}{2}}] + \mathrm{tr}[D^{-1}(\Omega - \Gamma^*\Lambda^{-1}\Gamma)].
\end{aligned}$$

This calculation depends on the existence of the inverse matrix Λ^{-1}. Now Λ is certainly positive definite if $A_{(k)}$ is positive definite, and Problem 14 asserts that $A_{(k)}$ is positive definite. It follows that Λ^{-1} not only exists but is positive definite as well. Furthermore, the matrix

$$D^{-\frac{1}{2}}(F - \Gamma^*\Lambda^{-1})\Lambda(F - \Gamma^*\Lambda^{-1})^* D^{-\frac{1}{2}}$$

is positive semidefinite and has a nonnegative trace. Hence, the maximum value of $Q(F, D \mid F_{(n)}, D_{(n)})$ with respect to F is attained at the point $F = \Gamma^*\Lambda^{-1}$, regardless of the value of D. In other words, the EM update of F is $F_{(n+1)} = \Gamma^*\Lambda^{-1}$. It should be stressed that Γ and Λ implicitly depend on the previous values $F_{(n)}$ and $D_{(n)}$. Once $F_{(n+1)}$ is determined, the equation

$$\begin{aligned} 0 &= \frac{\partial}{\partial d_i} Q(F, D \mid F_{(n)}, D_{(n)}) \\ &= -\frac{l}{2d_i} + \frac{1}{2d_i^2}(F\Lambda F^* - F\Gamma - \Gamma^* F^* + \Omega)_{ii} \end{aligned}$$

provides the update

$$d_{n+1,i} = \frac{1}{l}(F_{(n+1)}\Lambda F_{(n+1)}^* - F_{(n+1)}\Gamma - \Gamma^* F_{(n+1)}^* + \Omega)_{ii}.$$

One of the frustrating features of factor analysis is that the factor loading matrix F is not uniquely determined. To understand the source of the ambiguity, consider replacing F by FO, where O is a $q \times q$ orthogonal matrix. The distribution of each random vector $Y_{(k)}$ is normal with mean μ and variance matrix $FF^* + D$. If we substitute FO for F, then the variance $FOO^*F^* + D = FF^* + D$ remains the same. Another problem in factor analysis is the existence of more than one local maximum. Which one of these the EM algorithm converges to depends on its starting value [33].

7.6 Hidden Markov Chains

A hidden Markov chain incorporates both observed data and missing data. The missing data are the sequence of states visited by the chain; the observed data provide partial information about this sequence of states. Denote the sequence of visited states by $Z_1, \ldots, Z_n$ and the observation taken at epoch i when the chain is in state Z_i by $Y_i = y_i$. Baum's algorithms [4, 31] recursively compute the likelihood of the observed data

$$P = \Pr(Y_1 = y_1, \ldots, Y_n = y_n) \tag{7.6}$$

without actually enumerating all possible realizations $Z_1, \ldots, Z_n$. Baum's algorithms can be adapted to perform an EM search. The references [34, 83, 106] discuss several concrete examples of hidden Markov chains.

The likelihood (7.6) is constructed from three ingredients: (a) the initial distribution π at the first epoch of the chain, (b) the epoch-dependent transition probabilities $p_{ijk} = \Pr(Z_{i+1} = k \mid Z_i = j)$, and (c) the conditional densities $\phi_i(y_i \mid j) = \Pr(Y_i = y_i \mid Z_i = j)$. The dependence of the transition probability p_{ijk} on i allows the chain to be inhomogeneous over time and allows greater flexibility in modeling. Implicit in the definition of $\phi_i(y_i \mid j)$ are the assumptions that $Y_1, \ldots, Y_n$ are independent given $Z_1, \ldots, Z_n$ and that Y_i depends only on Z_i. For simplicity, we will assume that the Y_i are discretely distributed.

Baum's forward algorithm is based on recursively evaluating the joint probabilities

$$\alpha_i(j) = \Pr(Y_1 = y_1, \ldots, Y_{i-1} = y_{i-1}, Z_i = j).$$

At the first epoch, $\alpha_1(j) = \pi_j$ by definition; the obvious update to $\alpha_i(j)$ is

$$\alpha_{i+1}(k) = \sum_j \alpha_i(j)\phi_i(y_i \mid j)p_{ijk}. \tag{7.7}$$

The likelihood (7.6) can be recovered by computing the sum

$$P = \sum_j \alpha_n(j)\phi_n(y_n \mid j)$$

at the final epoch n.

In Baum's backward algorithm, we recursively evaluate the conditional probabilities

$$\beta_i(k) = \Pr(Y_{i+1} = y_{i+1}, \ldots, Y_n = y_n \mid Z_i = k),$$

starting by convention at $\beta_n(k) = 1$ for all k. The required update is clearly

$$\beta_i(j) = \sum_k p_{ijk}\phi_{i+1}(y_{i+1} \mid k)\beta_{i+1}(k). \tag{7.8}$$

In this instance, the likelihood is recovered at the first epoch by forming the sum $P = \sum_j \pi_j \phi_1(y_1 \mid j)\beta_1(j)$.

To see how Baum's algorithms interdigitate with the EM algorithm, we consider the problem of estimation with hidden multinomial trials. Suppose that the complete data involve N multinomial trials with success probability θ_k per trial for category k. Here N can be random or fixed. If N_k trials fall in category k, and there are l categories in all, then the complete data likelihood can be written as $c\prod_{k=1}^{l} \theta_k^{N_k}$, where c is an irrelevant constant. For example, in the hidden Markov chain setting, the hidden multinomial trials might involve (a) choice of the initial state, (b) choice of an observed outcome Y_i at the ith epoch given the hidden state j of the chain at that epoch, or (c) choice of the next state j given the current state i in a time-homogeneous chain. In the first case, the multinomial parameters are the π_i; in the last case, they are the common transition probabilities p_{ij}.

For hidden multinomial trials, the E step of the EM algorithm amounts to forming

$$Q(\theta \mid \theta_{(m)}) = \sum_{k=1}^{l} \mathrm{E}(N_k \mid Y = y, \theta_{(m)}) \ln \theta_k + \ln c,$$

where Y is the observed data and $\theta_{(m)}$ is the current parameter vector. The multinomial trials are hidden because only the function Y of the complete data is directly observed. Maximizing $Q(\theta \mid \theta_{(m)})$ subject to the constraints $\sum_{k=1}^{l} \theta_k = 1$ and $\theta_k \geq 0$ for all k is done as in Example 1.4.2. The resulting updates

$$\theta_{m+1,k} = \frac{\mathrm{E}(N_k \mid Y = y, \theta_{(m)})}{\mathrm{E}(N \mid Y = y, \theta_{(m)})}$$

are identical except that the expected counts $\mathrm{E}(N_k \mid Y = y, \theta_{(m)})$ may now be fractional numbers.

To flesh out these ideas, consider estimation of the initial distribution π at the first epoch of the chain. The number of trials N is the constant 1 for a single run of the Markov chain. Hence, $E(N \mid Y, \theta_{(m)}) = 1$. The conditional probability that the chain starts in state k given Y is just

$$\begin{aligned} \mathrm{E}(N_k \mid Y = y, \theta_{(m)}) &= \Pr(Z_1 = k \mid Y, \theta_{(m)}) \\ &= \frac{\pi_k \phi_1(y_1 \mid k)\beta_1(k)}{P}, \end{aligned}$$

where P is the likelihood of the observed data Y, $\beta_1(k)$ is a byproduct of the backward algorithm, and all computed quantities depend on the current parameter vector $\theta_{(m)}$. If there is more than one independent run of the chain, then these two conditional expectations are cumulated over the different runs.

If the conditional distribution of Y_i given $Z_i = j$ is multinomial with one trial and success probability θ_{ijk} for category k, then

$$\begin{aligned} \mathrm{E}(N \mid Y = y, \theta_{(m)}) &= \Pr(Z_i = j \mid Y, \theta_{(m)}) \\ &= \frac{\alpha_i(j)\phi_i(y_i \mid j)\beta_i(j)}{P} \\ \mathrm{E}(N_k \mid Y = y, \theta_{(m)}) &= \frac{1_{\{y_i = k\}}\alpha_i(j)\phi_i(y_i \mid j)\beta_i(j)}{P}, \end{aligned}$$

where $\phi_i(y_i \mid j) = \theta_{ijy_i}$. Again, if there is more than one independent run of the chain, then these two conditional expectations are cumulated over the different runs.

7.7 Problems

1. The entropy of a probability density $p(x)$ on R^n is defined by

$$-\int p(x) \ln p(x) dx. \tag{7.9}$$

Among all densities with a fixed mean vector $\mu = \int xp(x)dx$ and variance matrix $\Omega = \int (x-\mu)(x-\mu)^* p(x)dx$, prove that the multivariate normal has maximum entropy. (Hints: Apply Proposition 7.2.1 and formula (7.4).)

2. In statistical mechanics, entropy is employed to characterize the equilibrium distribution of many independently behaving particles. Let $p(x)$ be the probability density that a particle is found at position x in phase space R^n, and suppose that each position x is assigned an energy $u(x)$. If the average energy $U = \int u(x)p(x)dx$ per particle is fixed, then Nature chooses $p(x)$ to maximize entropy as defined in equation (7.9). Show that if constants α and β exist satisfying

$$\begin{aligned} \int \alpha e^{\beta u(x)} dx &= 1 \\ \int u(x) \alpha e^{\beta u(x)} dx &= U, \end{aligned}$$

then $p(x) = \alpha e^{\beta u(x)}$ does indeed maximize entropy subject to the average energy constraint. The density $p(x)$ is the celebrated Maxwell-Boltzmann density.

3. In the EM algorithm [25], suppose that the complete data X possesses a regular exponential density

$$f(x \mid \theta) = a(x) e^{b(\theta) + s(x)^* \theta}$$

relative to some measure ν. Prove that the unconditional mean of the sufficient statistic $s(X)$ is given by the negative gradient $-\nabla b(\theta)$ and that the EM update is characterized by the condition

$$\mathrm{E}[s(X) \mid Y, \theta_{(n)}] = -\nabla b(\theta_{(n+1)}).$$

4. Consider the data from *The London Times* [118] during the years 1910-1912 given in Table 7.1. The two columns labeled "Deaths i" refer to the number of deaths to women 80 years and older reported by day. The columns labeled "Frequency n_i" refer to the number of days with i deaths. A Poisson distribution gives a poor fit to these data, possibly because of different patterns of deaths in winter and summer. A mixture of two Poissons provides a much better fit. Under the Poisson admixture model, the likelihood of the observed data is

$$\prod_{i=0}^{9} \left[\alpha e^{-\mu_1} \frac{\mu_1^i}{i!} + (1-\alpha) e^{-\mu_2} \frac{\mu_2^i}{i!} \right]^{n_i},$$

where α is the admixture parameter and μ_1 and μ_2 are the means of the two Poisson distributions.

TABLE 7.1. Death Notices from *The London Times*

Deaths i	**Frequency** n_i	**Deaths** i	**Frequency** n_i
0	162	5	61
1	267	6	27
2	271	7	8
3	185	8	3
4	111	9	1

Formulate an EM algorithm for this model. Let $\theta = (\alpha, \mu_1, \mu_2)^*$ and

$$z_i(\theta) = \frac{\alpha e^{-\mu_1}\mu_1^i}{\alpha e^{-\mu_1}\mu_1^i + (1-\alpha)e^{-\mu_2}\mu_2^i}$$

be the posterior probability that a day with i deaths belongs to Poisson population 1. Show that the EM algorithm is given by

$$\begin{aligned}
\alpha_{m+1} &= \frac{\sum_i n_i z_i(\theta_{(m)})}{\sum_i n_i} \\
\mu_{m+1,1} &= \frac{\sum_i n_i i z_i(\theta_{(m)})}{\sum_i n_i z_i(\theta_{(m)})} \\
\mu_{m+1,2} &= \frac{\sum_i n_i i[1 - z_i(\theta_{(m)})]}{\sum_i n_i[1 - z_i(\theta_{(m)})]}.
\end{aligned}$$

From the initial estimates $\alpha_0 = 0.3$, $\mu_{01} = 1.$ and $\mu_{02} = 2.5$, compute via the EM algorithm the maximum likelihood estimates $\hat{\alpha} = 0.3599$, $\hat{\mu}_1 = 1.2561$, and $\hat{\mu}_2 = 2.6634$. Note how slowly the EM algorithm converges in this example.

5. Consider an i.i.d. sample drawn from a bivariate normal distribution with mean vector $\mu = (\mu_1, \mu_2)^*$ and variance matrix

$$\Omega = \begin{pmatrix} \sigma_1^2 & \sigma_{12} \\ \sigma_{12} & \sigma_2^2 \end{pmatrix}.$$

Suppose through some random accident that the first p observations are missing their first component, the next q observations are missing their second component, and the last r observations are complete. Design an EM algorithm to estimate the five mean and variance parameters, taking as complete data the original data before the accidental loss.

6. The standard linear regression model can be written in matrix notation as $X = A\beta + U$. Here X is the $r \times 1$ vector of dependent variables, A is the $r \times s$ design matrix, β is the $s \times 1$ vector of regression coefficients, and U is the $r \times 1$ normally distributed error vector with

mean $\mathbf{0}$ and variance $\sigma^2 I$. The dependent variables are right censored if for each i there is a constant c_i such that only $Y_i = \min\{c_i, X_i\}$ is observed. The EM algorithm offers a vehicle for estimating the parameter vector $\theta = (\beta^*, \sigma^2)^*$ in the presence of right censoring [25, 115]. Show that

$$\begin{aligned}
\beta_{(n+1)} &= (A^*A)^{-1}A^*\,\mathrm{E}(X \mid Y, \theta_{(n)}) \\
\sigma^2_{n+1} &= \frac{1}{r}\,\mathrm{E}[(X - A\beta_{(n+1)})^*(X - A\beta_{(n+1)}) \mid Y, \theta_{(n)}].
\end{aligned}$$

To compute the conditional expectations appearing in these formulas, let a_i be the ith row of A and define

$$H(v) = \frac{\frac{1}{\sqrt{2\pi}}e^{-\frac{v^2}{2}}}{\frac{1}{\sqrt{2\pi}}\int_v^\infty e^{-\frac{w^2}{2}}\,dw}.$$

For a censored observation $y_i = c_i < \infty$, prove that

$$\begin{aligned}
\mathrm{E}(X_i \mid Y_i = c_i, \theta_{(n)}) &= a_i\beta_{(n)} + \sigma_n H\Big(\frac{c_i - a_i\beta_{(n)}}{\sigma_n}\Big) \\
\mathrm{E}(X_i^2 \mid Y_i = c_i, \theta_{(n)}) &= (a_i\beta_{(n)})^2 + \sigma_n^2 \\
&\quad + \sigma_n(c_i + a_i\beta_{(n)})H\Big(\frac{c_i - a_i\beta_{(n)}}{\sigma_n}\Big).
\end{aligned}$$

Use these formulas to complete the specification of the EM algorithm.

7. Let $x_1, \ldots, x_m$ be i.i.d. observations drawn from a mixture of two normal densities with means μ_1 and μ_2 and common variance σ^2. These three parameters, together with the proportion α of observations taken from population 1, can be estimated by an EM algorithm. If

$$p_{ni} = \frac{\frac{\alpha_n}{\sqrt{2\pi\sigma_n^2}}e^{-\frac{(x_i-\mu_{n1})^2}{2\sigma_n^2}}}{\frac{\alpha_n}{\sqrt{2\pi\sigma_n^2}}e^{-\frac{(x_i-\mu_{n1})^2}{2\sigma_n^2}} + \frac{(1-\alpha_n)}{\sqrt{2\pi\sigma_n^2}}e^{-\frac{(x_i-\mu_{n2})^2}{2\sigma_n^2}}}$$

is the current posterior probability that observation i is taken from population 1, then derive the EM algorithm whose updates are

$$\begin{aligned}
\alpha_{n+1} &= \frac{1}{m}\sum_{i=1}^m p_{ni} \\
\mu_{n+1,1} &= \frac{\sum_{i=1}^m p_{ni}x_i}{\sum_{i=1}^m p_{ni}} \\
\mu_{n+1,2} &= \frac{\sum_{i=1}^m q_{ni}x_i}{\sum_{i=1}^m q_{ni}}
\end{aligned}$$

$$\sigma_{n+1}^2 = \frac{1}{m}\sum_{i=1}^m \left[p_{ni}(x_i - \mu_{n+1,1})^2 + q_{ni}(x_i - \mu_{n+1,2})^2\right],$$

where $q_{ni} = 1 - p_{ni}$.

8. In the transmission tomography model it is possible to approximate the solution of equation (7.3) to good accuracy in certain situations. Verify the expansion

$$\frac{1}{e^s - 1} = \frac{1}{s} - \frac{1}{2} + \frac{s}{12} + O(s^2).$$

Using the approximation $1/(e^s - 1) \approx 1/s - 1/2$ for $s = l_{ij}\theta_j$, show that

$$\theta_{n+1,j} = \frac{\sum_i (M_{ij} - N_{ij})}{\frac{1}{2}\sum_i (M_{ij} + N_{ij}) l_{ij}}$$

results. Can you motivate this result heuristically?

9. As an example of hidden binomial trials, consider a random sample of twin pairs. Let u of these pairs consist of male pairs, v consist of female pairs, and w consist of opposite sex pairs. A simple model to explain these data involves a random Bernoulli choice for each pair dictating whether it consists of identical or nonidentical twins. Suppose that identical twins occur with probability p and nonidentical twins with probability $1 - p$. Once the decision is made as to whether the twins are identical, then sexes are assigned to the twins. If the twins are identical, one assignment of sex is made. If the twins are nonidentical, then two independent assignments of sex are made. Suppose boys are chosen with probability q and girls with probability $1 - q$. Model these data as hidden binomial trials. Derive the EM algorithm for estimating p and q.

10. Chun Li has derived an alternative EM update for hidden multinomial trials. Let N denote the number of hidden trials, θ_i the probability of outcome i of k possible outcomes, and $L(\theta)$ the loglikelihood of the observed data Y. Derive the EM update

$$\theta_{n+1,i} = \theta_{ni} + \frac{\theta_{ni}}{\mathrm{E}(N \mid Y, \theta_{(n)})}\left[\frac{\partial}{\partial \theta_i}L(\theta_{(n)}) - \sum_{j=1}^k \theta_{nj}\frac{\partial}{\partial \theta_j}L(\theta_{(n)})\right].$$

11. In this problem you are asked to formulate models for hidden Poisson and exponential trials [120]. If the number of trials is N and the mean per trial is θ, then show that the EM update in the Poisson case is

$$\theta_{n+1} = \theta_n + \frac{\theta_n}{\mathrm{E}(N \mid Y, \theta_n)}\frac{d}{d\theta}L(\theta_n)$$

and in the exponential case is

$$\theta_{n+1} = \theta_n + \frac{\theta_n^2}{\mathrm{E}(N \mid Y, \theta_n)} \frac{d}{d\theta} L(\theta_n),$$

where $L(\theta)$ is the loglikelihood of the observed data Y.

12. Suppose light bulbs have an exponential lifetime with mean θ. Two experiments are conducted. In the first, the lifetimes $y_1, \ldots, y_m$ of m independent bulbs are observed. In the second, p independent bulbs are observed to burn out before time t, and q independent bulbs are observed to burn out after time t. In other words, the lifetimes in the second experiment are both left and right censored. Construct an EM algorithm for finding the maximum likelihood estimate of θ [45].

13. Suppose X and Y are independent random vectors. Show that the Kronecker product $X \otimes Y$ has mean vector and variance matrix

$$\begin{aligned} \mathrm{E}(X \otimes Y) &= \mathrm{E}(X) \otimes \mathrm{E}(Y) \\ \mathrm{Var}(X \otimes Y) &= \mathrm{Var}(X) \otimes \mathrm{Var}(Y) + [\mathrm{E}(X)\,\mathrm{E}(X)^*] \otimes \mathrm{Var}(Y) \\ &\quad + \mathrm{Var}(X) \otimes [\mathrm{E}(Y)\,\mathrm{E}(Y)^*]. \end{aligned}$$

14. Suppose that Σ is a positive definite matrix. Prove that the matrix $I - F^*(FF^* + \Sigma)^{-1}F$ is also positive definite. This result is used in the derivation of the EM algorithm in Section 7.5. (Hints: For readers familiar with the sweep operator of computational statistics, the simplest proof relies on applying Propositions 7.5.2 and 7.5.3 of [82].)

15. In the hidden Markov chain model, suppose that the chain is time homogeneous with transition probabilities p_{ij}. Derive an EM algorithm for estimating the p_{ij} from one or more independent runs of the chain.

8

Newton's Method

8.1 Introduction

The MM and EM algorithms are hardly the only methods of optimization. Newton's method is better known and more widely applied. Despite its defects, Newton's method is the gold standard for speed of convergence and forms the basis of most modern optimization algorithms. Its many variants seek to retain its fast convergence while taming its defects. They all revolve around the core idea of locally approximating the objective function by a strictly convex quadratic function. At each iteration the quadratic approximation is optimized. Safeguards are introduced to keep the iterates from veering toward irrelevant stationary points.

Statisticians are among the most avid consumers of optimization techniques. Statistics, like other scientific disciplines, has a special vocabulary. We will meet some of that vocabulary in this chapter as we discuss optimization methods important in computational statistics. Thus, we will take up Fisher's scoring algorithm and the Gauss-Newton method of nonlinear least squares. We have already encountered likelihood functions and the device of passing to loglikelihoods. In statistics, the gradient of the loglikelihood is called the score, and the negative of the second differential is called the observed information. One major advantage of maximizing the loglikelihood rather than the likelihood is that the loglikelihood, score, and observed information are all additive functions of independent observations.

8.2 Newton's Method and Root Finding

One of the virtues of Newton's method is that it is a root-finding technique as well as an optimization technique. Consider a function $f(x)$ mapping R^n into R^n, and suppose a root of $f(x) = \mathbf{0}$ occurs at y. If the slope matrix in the expansion

$$\begin{aligned} \mathbf{0} - f(x) &= f(y) - f(x) \\ &= s(y, x)(y - x) \end{aligned}$$

is invertible, then we can solve for y as

$$y = x - s(y, x)^{-1} f(x).$$

In practice, y is unknown and the slope $s(y, x)$ is unavailable. However, if y is close to x, then $s(y, x)$ should be close to $df(x)$. Thus, Newton's method iterates according to

$$x_{(m+1)} = x_{(m)} - df(x_{(m)})^{-1} f(x_{(m)}). \tag{8.1}$$

Example 8.2.1 *Division without Dividing*

Forming the reciprocal of a number $a > 0$ is equivalent to solving for a root of the equation $f(x) = a - x^{-1}$. Newton's method (8.1) iterates according to

$$\begin{aligned} x_{m+1} &= x_m - \frac{a - x_m^{-1}}{x_m^{-2}} \\ &= x_m(2 - a x_m), \end{aligned}$$

which involves multiplication and subtraction but no division. If x_{m+1} is to be positive, then x_m must lie on the interval $(0, 2/a)$. If x_m does indeed reside there, then x_{m+1} will reside on the shorter interval $(0, 1/a)$ because the quadratic $x(2 - ax)$ attains its maximum of $1/a$ at $x = 1/a$. Furthermore, $x_{m+1} > x_m$ if and only if $2 - ax_m > 1$, and this latter inequality holds if and only if $x_m < 1/a$. Thus, starting on $(0, 1/a)$, the iterates x_m monotonically increase to their limit $1/a$. Starting on $[1/a, 2/a)$, the first iterate satisfies $x_1 \leq 1/a$, and subsequent iterates monotonically increase to $1/a$. Finally, starting outside $(0, 2/a)$ leads either to fixation at 0 or divergence to $-\infty$. ■

Example 8.2.2 *Extraction of nth Roots*

Newton's method can be used to extract square roots, cube roots, and so forth. Consider the function $f(x) = x^n - a$ for some integer $n > 1$ and $a > 0$. Newton's method amounts to the iteration scheme

$$x_{m+1} = x_m - \frac{x_m^n - a}{n x_m^{n-1}} = \frac{1}{n}\left[(n-1)x_m + \frac{a}{x_m^{n-1}}\right]. \tag{8.2}$$

TABLE 8.1. Newton's Iterates for $x^4 - x^2$

m	x_m	x_m	x_m
0	-0.74710	-0.66710	-0.500000
1	-2.16581	1.01669	-0.125000
2	-1.68896	1.00066	-0.061491
3	-1.35646	1.00000	-0.030628
4	-1.14388	1.00000	-0.015300
5	-1.03477	1.00000	-0.007648
6	-1.00270	1.00000	-0.003823
7	-1.00002	1.00000	-0.001911
8	-1.00000	1.00000	-0.000955
9	-1.00000	1.00000	-0.000477
10	-1.00000	1.00000	-0.000238

This sequence converges to $\sqrt[n]{a}$ regardless of the starting point $x_0 > 0$. To demonstrate this fact, we first note that the right-hand side of equation (8.2) is the arithmetic mean of $n-1$ copies of the number x_m and a/x_m^{n-1}. Because the arithmetic mean exceeds the geometric mean $\sqrt[n]{a}$, it follows that $x_m \geq \sqrt[n]{a}$ for all $m \geq 1$. Given this inequality, we have $a/x_m^{n-1} \leq x_m$. Again viewing equation (8.2) as a weighted average of x_m and the ratio $a/x_m^{n-1} \leq x_m$, it follows that $x_{m+1} \leq x_m$ for all $m \geq 1$. Hence, the sequence $x_1, x_2, \ldots$ is bounded below and is monotone decreasing. By continuity, its limit is $\sqrt[n]{a}$. ∎

Example 8.2.3 *Sensitivity to Initial Conditions*

In contrast to the previous two well-behaved examples, finding a root of the polynomial $f(x) = x^4 - x^2$ is more problematic. These roots are clearly -1, 0, and 1. We anticipate trouble when $f'(x) = 4x^3 - 2x = 0$ at the points $-1/\sqrt{2}$, 0, and $1/\sqrt{2}$. Consider initial points near $-1/\sqrt{2}$. Just to the left of $-1/\sqrt{2}$, Newton's method converges to -1. For a narrow zone just to the right of $-1/\sqrt{2}$, it converges to 1, and beyond this zone but to the left of $1/\sqrt{2}$, it converges to 0. Table 8.1 gives three typical examples of this extreme sensitivity to initial conditions. The slower convergence to the middle root 0 is hardly surprising given that $f'(0) = 0$. It appears that the discrepancy $x_m - 0$ roughly halves at each iteration. Problem 2 clarifies this behavior. ∎

Example 8.2.4 *Secant Method*

There are several ways of estimating the derivative $f'(x_{(m)})$ appearing in Newton's formula (8.1). The secant method approximates $f'(x_{(m)})$ by the slope $s(x_{(m-1)}, x_{(m)})$ using the canonical slope function

$$s(y,x) = \frac{f(y)-f(x)}{y-x}$$

in one dimension. This produces the secant update

$$x_{m+1} = x_m - \frac{f(x_m)(x_{m-1}-x_m)}{f(x_{m-1})-f(x_m)}, \tag{8.3}$$

which is the prototype for the quasi-Newton updates treated in later chapters. While the secant method avoids computation of derivatives, it typically takes more iterations to converge than Newton's method. Safeguards must also be put into place to ensure its reliability. ■

Example 8.2.5 *Newton's Method of Matrix Inversion*

Newton's method for finding the reciprocal of a number can be generalized to compute the inverse of a matrix [67]. Consider the iteration scheme

$$B_{(m+1)} = 2B_{(m)} - B_{(m)}AB_{(m)}$$

for some fixed $n \times n$ matrix A. Rearranging this equation yields

$$A^{-1} - B_{(m+1)} = (A^{-1} - B_{(m)})A(A^{-1} - B_{(m)}),$$

which implies that

$$\|A^{-1} - B_{(m+1)}\| \leq \|A\| \cdot \|A^{-1} - B_{(m)}\|^2$$

for every matrix norm. It follows that the sequence $B_{(m)}$ converges at a quadratic rate to A^{-1} if $B_{(1)}$ is sufficiently close to A^{-1}. ■

8.3 Newton's Method and Optimization

Suppose we want to minimize the real-valued function $f(x)$ defined on an open set $S \subset \mathsf{R}^n$. Assuming that $f(x)$ is twice differentiable, the expansion

$$f(y) = f(x) + df(x)(y-x) + \frac{1}{2}(y-x)^* s^2(y,x)(y-x)$$

suggests that we substitute $d^2f(x)$ for the second slope $s^2(y,x)$ and approximate $f(y)$ by the resulting quadratic. If we take this approximation seriously, then we can solve for its minimum point y as

$$y = x - d^2f(x)^{-1}\nabla f(x).$$

In Newton's method we iterate according to

$$x_{(m+1)} = x_{(m)} - d^2 f(x_{(m)})^{-1} \nabla f(x_{(m)}). \qquad (8.4)$$

It should come as no surprise that algorithm (8.4) coincides with the earlier version of Newton's method seeking a root of $\nabla f(x)$. For this reason, any stationary point x of $f(x)$ is a fixed point of algorithm (8.4).

There are two potential problems with Newton's method. First, it may be expensive computationally to evaluate or invert $d^2 f(x)$. Second, far from the minimum, Newton's method is equally happy to head uphill or down. In other words, Newton's method is not a descent algorithm in the sense that $f(x_{(m+1)}) < f(x_{(m)})$. This second defect can be remedied by modifying the Newton increment so that it is a partial step in a descent direction. A descent direction v at the point x satisfies the inequality $df(x)v < 0$. The formula

$$\lim_{t \downarrow 0} \frac{f(x + tv) - f(x)}{t} = df(x)v$$

for the forward directional derivative shows that $f(x+tv) < f(x)$ for $t > 0$ sufficiently small. The key to generating a descent direction is to define $v = -H^{-1}\nabla f(x)$ using a positive definite matrix H. Here we assume that x is not a stationary point and recall the fact that the inverse of a positive definite matrix is positive definite.

The necessary modifications of Newton's method to achieve a descent algorithm are now clear. We simply replace $d^2 f(x_{(m)})$ by a positive definite approximating matrix $H_{(m)}$ and take a sufficiently short step in the direction $\Delta x_{(m)} = -H_{(m)}^{-1} \nabla f(x_{(m)})$. If we believe that the proposed increment is reasonable, then we will be reluctant to shrink $\Delta x_{(m)}$ too much. This suggests backtracking, the simplest form of which is step halving. In step halving, if the initial increment $\Delta x_{(m)}$ does not produce a decrease in $f(x)$, then try $\Delta x_{(m)}/2$. If $\Delta x_{(m)}/2$ fails, then try $\Delta x_{(m)}/4$, and so forth. We will meet more sophisticated backtracking schemes later. Note at this juncture we have said nothing about how well $H_{(m)}$ approximates $d^2 f(x_{(m)})$. The quality of this approximation obviously affects the rate of convergence toward any local minimum.

If we minimize $f(x)$ subject to the linear equality constraints $Vx = d$, then minimization of the approximating quadratic can be accomplished as indicated in Example 4.2.5 of Chapter 4. Because

$$V(x_{(m+1)} - x_{(m)}) = \mathbf{0},$$

the revised increment $\Delta x_{(m)} = x_{(m+1)} - x_{(m)}$ is

$$\Delta x_{(m)} = -\left[H_{(m)}^{-1} - H_{(m)}^{-1} V^* (V H_{(m)}^{-1} V^*)^{-1} V H_{(m)}^{-1}\right] \nabla f(x_{(m)}). \qquad (8.5)$$

This can be viewed as the projection of the unconstrained increment onto the null space of V. Problem 4 shows that step halving also works for the projected increment.

8.4 MM Gradient Algorithm

Often it is impossible to solve the optimization step of the MM algorithm exactly. If $f(x)$ is the objective function and $g(x \mid x_{(m)})$ minorizes or majorizes $f(x)$ at $x_{(m)}$, then Newton's method can be applied to optimize $g(x \mid x_{(m)})$. As we shall see later, one step of Newton's method preserves the overall rate of convergence of the MM algorithm. Thus, the MM gradient algorithm iterates according to

$$\begin{aligned} x_{(m+1)} &= x_{(m)} - d^2 g(x_{(m)} \mid x_{(m)})^{-1} \nabla g(x_{(m)} \mid x_{(m)}) \\ &= x_{(m)} - d^2 g(x_{(m)} \mid x_{(m)})^{-1} \nabla f(x_{(m)}). \end{aligned}$$

Here derivatives are taken with respect to the left argument of $g(x \mid x_{(m)})$. Substitution of $\nabla f(x_{(m)})$ for $\nabla g(x_{(m)} \mid x_{(m)})$ can be justified by noting that the difference $f(x) - g(x \mid x_{(m)})$ attains either its minimum or maximum at $x = x_{(m)}$. If $x_{(m)}$ is interior to the domain of $f(x)$, then the gradient of $f(x) - g(x \mid x_{(m)})$ vanishes there. In most practical examples, the surrogate function $g(x \mid x_{(m)})$ is either convex or concave, and its second differential $d^2 g(x_{(m)} \mid x_{(m)})$ needs no adjustment to give a descent or ascent algorithm.

Example 8.4.1 *Newton's Method in Transmission Tomography*

In the transmission tomography model of Chapter 6, the surrogate function $g(\theta \mid \theta_{(m)})$ of equation (6.9) minorizes the loglikelihood $L(\theta)$ in the absence of a smoothing prior. Differentiating $g(\theta \mid \theta_{(m)})$ with respect to θ_j gives the transcendental equation

$$0 = \sum_i l_{ij} \Big[d_i e^{-l_i^* \theta_{(m)} \theta_j / \theta_{mj}} - y_i \Big].$$

One step of Newton's method starting at $\theta_j = \theta_{mj}$ produces the next iterate

$$\begin{aligned} \theta_{m+1,j} &= \theta_{mj} + \frac{\theta_{mj} \sum_i l_{ij} (d_i e^{-l_i^* \theta_{(m)}} - y_i)}{\sum_i l_{ij} l_i^* \theta_{(m)} d_i e^{-l_i^* \theta_{(m)}}} \\ &= \theta_{mj} \frac{\sum_i l_{ij} [d_i e^{-l_i^* \theta_{(m)}} (1 + l_i^* \theta_{(m)}) - y_i]}{\sum_i l_{ij} l_i^* \theta_{(m)} d_i e^{-l_i^* \theta_{(m)}}}. \end{aligned}$$

This step typically increases $L(\theta)$. ■

Example 8.4.2 *Estimation with the Dirichlet Distribution*

As another example, consider parameter estimation for the Dirichlet distribution [76]. This distribution has probability density

$$\frac{\Gamma(\sum_{i=1}^j \theta_i)}{\prod_{i=1}^j \Gamma(\theta_i)} \prod_{i=1}^j y_i^{\theta_i - 1} \tag{8.6}$$

on the simplex $\{y = (y_1, \ldots, y_j)^* : y_1 > 0, \ldots, y_j > 0, \sum_{i=1}^j y_i = 1\}$ endowed with the uniform measure. The Dirichlet distribution is used to represent random proportions.

If $y_{(1)}, \ldots, y_{(l)}$ are randomly sampled vectors from the Dirichlet distribution, then their loglikelihood is

$$L(\theta) = l \ln \Gamma\Big(\sum_{i=1}^j \theta_i\Big) - l \sum_{i=1}^j \ln \Gamma(\theta_i) + \sum_{k=1}^l \sum_{i=1}^j (\theta_i - 1) \ln y_{ki}.$$

Except for the first term on the right, the parameters are separated. Fortunately as demonstrated in Example 5.3.10, the function $\ln \Gamma(t)$ is convex. Denoting its derivative by $\psi(t)$, we exploit the minorization

$$\ln \Gamma\Big(\sum_{i=1}^j \theta_i\Big) \geq \ln \Gamma\Big(\sum_{i=1}^j \theta_{mi}\Big) + \psi\Big(\sum_{i=1}^j \theta_{mi}\Big) \sum_{i=1}^j (\theta_i - \theta_{mi})$$

and create the surrogate function

$$\begin{aligned} g(\theta \mid \theta_{(m)}) &= l \ln \Gamma\Big(\sum_{i=1}^j \theta_{mi}\Big) + l \psi\Big(\sum_{i=1}^j \theta_{mi}\Big) \sum_{i=1}^j (\theta_i - \theta_{mi}) \\ &\quad - l \sum_{i=1}^j \ln \Gamma(\theta_i) + \sum_{k=1}^l \sum_{i=1}^j (\theta_i - 1) \ln y_{ki}. \end{aligned}$$

Owing to the presence of the terms $\ln \Gamma(\theta_i)$, the maximization step is intractable. However, the MM gradient algorithm can be readily implemented because the parameters are now separated and the functions $\psi(t)$ and $\psi'(t)$ are easily computed as suggested in Problem 5. The whole process is carried out in the references [81, 99] on actual data. ■

8.5 Ad Hoc Approximations of $d^2 f(\theta)$

In minimization problems, we have emphasized the importance of approximating $d^2 f(x)$ by a positive definite matrix. Two key ideas drive the process of approximation. One is the recognition that outer product matrices are positive semidefinite. Another is a feel for when terms are small on average. Usually this involves comparison of random variables and their means.

For example, consider the problem of least squares estimation with nonlinear regression functions. Let us formulate the problem slightly more generally as one of minimizing the sum of squares

$$f(\theta) = \frac{1}{2} \sum_{i=1}^j w_i [y_i - \mu_i(\theta)]^2$$

involving weights $w_i > 0$ and observations y_i for each case i. Here y_i is a realization of a random variable Y_i with mean $\mu_i(\theta)$. In linear regression, $\mu_i(\theta) = \sum_k x_{ik}\theta_k$. To implement Newton's method, we need

$$\begin{aligned} \nabla f(\theta) &= -\sum_{i=1}^{j} w_i[y_i - \mu_i(\theta)]\nabla\mu_i(\theta) \\ d^2 f(\theta) &= \sum_{i=1}^{j} w_i \nabla\mu_i(\theta) d\mu_i(\theta) - \sum_{i=1}^{j} w_i[y_i - \mu_i(\theta)] d^2\mu_i(\theta). \end{aligned} \tag{8.7}$$

In the Gauss-Newton algorithm, we approximate

$$d^2 f(\theta) \approx \sum_{i=1}^{j} w_i \nabla\mu_i(\theta) d\mu_i(\theta)$$

on the rationale that either the weighted residuals $w_i[x_i - \mu_i(\theta)]$ are small or the regression functions $\mu_i(\theta)$ are nearly linear. In both instances, the Gauss-Newton algorithm shares the fast convergence of Newton's method.

Maximum likelihood estimation with the Poisson distribution furnishes another example. Here the count data $y_1, \ldots, y_j$ have loglikelihood, score, and negative observed information

$$\begin{aligned} L(\theta) &= \sum_{i=1}^{j} [y_i \ln\mu_i(\theta) - \mu_i(\theta) - \ln y_i!] \\ \nabla L(\theta) &= \sum_{i=1}^{j} \Big[\frac{y_i}{\mu_i(\theta)}\nabla\mu_i(\theta) - \nabla\mu_i(\theta)\Big] \\ d^2 L(\theta) &= \sum_{i=1}^{j} \Big[-\frac{y_i}{\mu_i(\theta)^2}\nabla\mu_i(\theta) d\mu_i(\theta) + \frac{y_i}{\mu_i(\theta)} d^2\mu_i(\theta) - d^2\mu_i(\theta)\Big]. \end{aligned}$$

In view of the fact that $\mathrm{E}(y_i) = \mu_i(\theta)$, the negative semidefinite approximations

$$\begin{aligned} d^2 L(\theta) &\approx -\sum_{i=1}^{j} \frac{y_i}{\mu_i(\theta)^2}\nabla\mu_i(\theta) d\mu_i(\theta) \\ &\approx -\sum_{i=1}^{j} \frac{1}{\mu_i(\theta)}\nabla\mu_i(\theta) d\mu_i(\theta) \end{aligned}$$

are reasonable. The second of these leads to the scoring algorithm discussed in the next section.

The exponential distribution offers a third illustration. Now the data have means $\mathrm{E}(y_i) = 1/\mu_i(\theta)$. The loglikelihood

$$L(\theta) = \sum_{i=1}^{j} [\ln\mu_i(\theta) - y_i\mu_i(\theta)]$$

yields the score and negative observed information

$$\begin{aligned} \nabla L(\theta) &= \sum_{i=1}^{j} \Big[\frac{1}{\mu_i(\theta)} \nabla \mu_i(\theta) - y_i \nabla \mu_i(\theta)\Big] \\ d^2 L(\theta) &= \sum_{i=1}^{j} \Big[-\frac{1}{\mu_i(\theta)^2} \nabla \mu_i(\theta) d\mu_i(\theta) + \frac{1}{\mu_i(\theta)} d^2 \mu_i(\theta) - y_i d^2 \mu_i(\theta)\Big]. \end{aligned}$$

Replacing observations by their means suggests the approximation

$$d^2 L(\theta) \approx -\sum_{i=1}^{j} \frac{1}{\mu_i(\theta)^2} \nabla \mu_i(\theta) d\mu_i(\theta)$$

made in the scoring algorithm.

Our final example involves maximum likelihood estimation with the multinomial distribution. The observations $y_1, \ldots, y_j$ are now cell counts over n independent trials. Cell i is assigned probability $p_i(\theta)$ and averages a total of $np_i(\theta)$ counts. The loglikelihood, score, and negative observed information amount to

$$\begin{aligned} L(\theta) &= \sum_{i=1}^{j} y_i \ln p_i(\theta) \\ \nabla L(\theta) &= \sum_{i=1}^{j} \frac{y_i}{p_i(\theta)} \nabla p_i(\theta) \\ d^2 L(\theta) &= \sum_{i=1}^{j} \Big[-\frac{y_i}{p_i(\theta)^2} \nabla p_i(\theta) dp_i(\theta) + \frac{y_i}{p_i(\theta)} d^2 p_i(\theta)\Big]. \end{aligned}$$

In light of the identity $\mathrm{E}(y_i) = np_i(\theta)$, the approximation

$$\sum_{i=1}^{j} \frac{y_i}{p_i(\theta)} d^2 p_i(\theta) \approx n \sum_{i=1}^{j} d^2 p_i(\theta) = d^2 n = 0$$

is reasonable. This suggests the further negative semidefinite approximations

$$\begin{aligned} d^2 L(\theta) &\approx -\sum_{i=1}^{j} \frac{y_i}{p_i(\theta)^2} \nabla p_i(\theta) dp_i(\theta) \\ &\approx -n \sum_{i=1}^{j} \frac{1}{p_i(\theta)} \nabla p_i(\theta) dp_i(\theta), \end{aligned}$$

the second of which coincides with the scoring algorithm.

8.6 Scoring

As we just have witnessed, one can approximate the observed information in a variety of ways. If $L(\theta)$ is the loglikelihood, the scoring algorithm replaces the observed information $-d^2L(\theta)$ by the expected information $J(\theta) = \mathrm{E}[-d^2L(\theta)]$. In the examples we have studied, $J(\theta)$ appears to involve first derivatives rather than second derivatives. In fact, one can also represent $J(\theta)$ as the covariance matrix $\mathrm{Var}[\nabla L(\theta)]$. This incidentally shows that $J(\theta)$ is positive semidefinite and is a good replacement for $-d^2L(\theta)$ in Newton's method. An extra dividend of scoring is that the inverse matrix $J(\hat{\theta})^{-1}$ immediately supplies the asymptotic variances and covariances of the maximum likelihood estimate $\hat{\theta}$ [107]. Scoring shares this benefit with Newton's method since the observed information is under natural assumptions asymptotically equivalent to the expected information.

To prove that $J(\theta) = \mathrm{Var}[\nabla L(\theta)]$, suppose the data has density $f(y \mid \theta)$ relative to some measure ν, which is usually ordinary volume measure or a discrete counting measure. We first note that the score conveniently has vanishing expectation because

$$\begin{aligned}
\mathrm{E}[\nabla L(\theta)] &= \int \frac{\nabla f(y \mid \theta)}{f(y \mid \theta)} f(y \mid \theta)\, d\nu(y) \\
&= \nabla \int f(y \mid \theta)\, d\nu(y)
\end{aligned}$$

and $\int f(y,\theta) d\nu(y) = 1$. Here the interchange of differentiation and expectation must be proved, but we will not stop to do so. See the references [87, 107]. The formal calculation

$$\begin{aligned}
\mathrm{E}[-d^2L(\theta)] &= -\int \left[\frac{d^2 f(y \mid \theta)}{f(y \mid \theta)} - \frac{\nabla f(y \mid \theta) df(y \mid \theta)}{f(y \mid \theta)^2}\right] f(y \mid \theta)\, d\nu(y) \\
&= -d^2 \int f(y \mid \theta)\, d\nu(y) \\
&\quad + \int \nabla L(\theta) dL(\theta) f(y \mid \theta)\, d\nu(y) \\
&= -\mathbf{0} + \mathrm{E}[\nabla L(\theta) dL(\theta)]
\end{aligned}$$

then completes the verification.

Example 8.6.1 *Scoring and the Gauss-Newton Algorithm*

Armed with our better understanding of scoring, let us revisit nonlinear regression. Suppose that the j independent observations $y_1, \ldots, y_j$ are normally distributed with means $\mu_i(\theta)$ and variances σ^2/w_i, where the w_i are known constants. To estimate the mean parameter vector θ and the variance parameter σ^2 by scoring, we first write the loglikelihood up to a constant as the function

$$\begin{aligned} L(\phi) &= -\frac{j}{2}\ln\sigma^2 - \frac{1}{2\sigma^2}\sum_{i=1}^{j} w_i[y_i - \mu_i(\theta)]^2 \\ &= -\frac{j}{2}\ln\sigma^2 - \frac{f(\theta)}{\sigma^2} \end{aligned}$$

of the parameters $\phi = (\theta^*, \sigma^2)^*$.

Straightforward differentiation yields the score

$$\nabla L(\phi) = \begin{pmatrix} \frac{1}{\sigma^2}\sum_{i=1}^{j} w_i[y_i - \mu_i(\theta)]\nabla\mu_i(\theta) \\ -\frac{j}{2\sigma^2} + \frac{1}{2\sigma^4}\sum_{i=1}^{j} w_i[y_i - \mu_i(\theta)]^2 \end{pmatrix}.$$

To derive the expected information

$$J(\phi) = \begin{pmatrix} \frac{1}{\sigma^2}\sum_{i=1}^{j} w_i\nabla\mu_i(\theta)d\mu_i(\theta) & 0 \\ 0 & \frac{j}{2\sigma^4} \end{pmatrix},$$

we note that the observed information matrix can be written as the symmetric block matrix

$$-d^2L(\phi) = \begin{pmatrix} H_{\theta\theta} & H_{\theta\sigma^2} \\ H_{\sigma^2\theta} & H_{\sigma^2\sigma^2} \end{pmatrix}.$$

The upper-left block $H_{\theta\theta}$ equals $d^2f(\theta)/\sigma^2$ with $d^2f(\theta)$ given by equation (8.7). The displayed value of the expectation $\mathrm{E}(H_{\theta\theta})$ follows directly from the identity $\mathrm{E}[y_i - \mu_i(\theta)] = 0$. The upper-right block $H_{\theta\sigma^2}$ amounts to $\nabla f(\theta)/\sigma^4$, and its expectation vanishes because $\mathrm{E}[y_i - \mu_i(\theta)] = 0$. Finally, the lower-right block $H_{\sigma^2\sigma^2}$ equals

$$-\frac{j}{2\sigma^4} + \frac{1}{\sigma^6}\sum_{i=1}^{j} w_i[y_i - \mu_i(\theta)]^2.$$

Its expectation

$$\begin{aligned} \mathrm{E}(H_{\sigma^2\sigma^2}) &= -\frac{j}{2\sigma^4} + \frac{1}{\sigma^6}\sum_{i=1}^{j} w_i\,\mathrm{E}\{[y_i - \mu_i(\theta)]^2\} \\ &= \frac{j}{2\sigma^4} \end{aligned}$$

because $\mathrm{Var}(y_i) = \sigma^2/w_i$. Readers experienced in calculating variances and covariances can verify the blocks of $J(\theta)$ by forming $\mathrm{Var}[\nabla L(\theta)]$.

In any event, scoring updates θ by

$$\begin{aligned} &\theta_{(m+1)} \qquad\qquad (8.8) \\ &= \theta_{(m)} + \Big[\sum_{i=1}^{j} w_i\nabla\mu_i(\theta_{(m)})d\mu(\theta_{(m)})\Big]^{-1}\sum_{i=1}^{j} w_i[y_i - \mu_i(\theta_{(m)})]\nabla\mu_i(\theta_{(m)}) \end{aligned}$$

and σ^2 by

$$\sigma^2_{m+1} = \frac{1}{j}\sum_{i=1}^{j} w_i[y_i - \mu_i(\theta_{(m)})]^2.$$

The scoring algorithm (8.8) for θ amounts to nothing more than the Gauss-Newton algorithm. The Gauss-Newton updates can be carried out blithely neglecting the updates of σ^2. ∎.

The score and expected information simplify considerably for exponential families of densities [10, 17, 51, 71, 101]. Such densities take the form

$$f(y \mid \theta) = g(y)e^{\beta(\theta)+h(y)^*\gamma(\theta)} \tag{8.9}$$

relative to some measure ν. Most of the distributional families commonly encountered in statistics are exponential families. The normal, Poisson, binomial, negative binomial, gamma, beta, and multinomial families are prime examples. The score and expected information can be expressed succinctly in terms of the mean vector $\mu(\theta) = \mathrm{E}[h(y)]$ and the variance matrix $\Sigma(\theta) = \mathrm{Var}[h(y)]$ of the sufficient statistic $h(y)$. Our point of departure in deriving these quantities is the identity

$$dL(\theta) = d\beta(\theta) + h(y)^* d\gamma(\theta). \tag{8.10}$$

If $\gamma(\theta)$ is linear in θ, then $J(\theta) = -d^2L(\theta) = -d^2\beta(\theta)$, and scoring coincides with Newton's method. If in addition $J(\theta)$ is positive definite, then $L(\theta)$ is strictly concave and possesses at most a single local maximum, which is necessarily the global maximum.

For an exponential family, the fact that $\mathrm{E}[\nabla L(\theta)] = \mathbf{0}$ can be restated as

$$d\beta(\theta) + \mu(\theta)^* d\gamma(\theta) = \mathbf{0}^*. \tag{8.11}$$

Subtracting equation (8.11) from equation (8.10) yields the alternative representation

$$dL(\theta) = [h(y) - \mu(\theta)]^* d\gamma(\theta) \tag{8.12}$$

of the first differential. This representation implies that the expected information is

$$\begin{aligned} J(\theta) &= \mathrm{Var}[\nabla L(\theta)] \\ &= d\gamma(\theta)^* \Sigma(\theta) d\gamma(\theta). \end{aligned} \tag{8.13}$$

To eliminate $d\gamma(\theta)$ in equations (8.12) and (8.13), note that

$$d\mu(\theta) = \int h(y) df(y \mid \theta)\, d\nu(y)$$

$$
\begin{aligned}
&= \int h(y) dL(\theta) f(y \mid \theta)\, d\nu(y) \\
&= \int h(y)[h(y) - \mu(\theta)]^* \, d\gamma(\theta) f(y \mid \theta)\, d\nu(y) \\
&= \Sigma(\theta) d\gamma(\theta).
\end{aligned}
$$

When $\Sigma(\theta)$ is invertible, this calculation implies $d\gamma(\theta) = \Sigma(\theta)^{-1} d\mu(\theta)$, which in view of (8.12) and (8.13) yields

$$
\begin{aligned}
dL(\theta) &= [h(y) - \mu(\theta)]^* \Sigma(\theta)^{-1} d\mu(\theta) \\
J(\theta) &= d\mu(\theta)^* \Sigma(\theta)^{-1} d\mu(\theta).
\end{aligned} \tag{8.14}
$$

One can verify these formulas directly for the normal, Poisson, exponential, and multinomial distributions studied in the previous section. In each instance the sufficient statistic for case i is just y_i.

Example 8.6.2 *Scoring for the ABO Example*

In the ABO allele frequency estimation problem studied in Chapter 6, scoring can be implemented by taking as basic parameters p_A and p_B and expressing $p_O = 1 - p_A - p_B$. Scoring then leads to the same maximum likelihood point $(\hat{p}_A, \hat{p}_B, \hat{p}_O) = (.2136, .0501, .7363)$ as the MM algorithm. The quicker convergence of scoring here—four iterations as opposed to five starting from $(.3, .2, .5)$—is often more dramatic in other problems. Scoring also has the advantage over MM of immediately providing asymptotic standard deviations of the parameter estimates. These are $(.0135, .0068, .0145)$ for the estimates $(\hat{p}_A, \hat{p}_B, \hat{p}_O)$. ■

Example 8.6.3 *Generalized Linear Models*

The generalized linear model [101] deals with exponential families (8.9) in which the sufficient statistic $h(y)$ is y and the mean $\mu(\theta)$ of y completely determines the distribution of y. In many applications it is natural to postulate that $\mu(\theta) = q(x^*\theta)$ is a monotone function q of some linear combination of known covariates x. The inverse of q is called the link function. In this setting, $d\mu(\theta) = q'(x^*\theta)x^*$. It follows from equation (8.14) that if $y_1, \ldots, y_j$ are independent observations with corresponding covariate vectors $x_{(1)}, \ldots, x_{(j)}$, then the score and expected information can be written as

$$
\begin{aligned}
\nabla L(\theta) &= \sum_{i=1}^{j} \frac{y_i - \mu_i(\theta)}{\sigma_i^2(\theta)} q'(x_{(i)}^* \theta) x_{(i)} \\
J(\theta) &= \sum_{i=1}^{j} \frac{1}{\sigma_i^2(\theta)} q'(x_{(i)}^* \theta)^2 x_{(i)} x_{(i)}^*,
\end{aligned}
$$

where $\sigma_i^2(\theta) = \operatorname{Var}(y_i)$.

TABLE 8.2. AIDS Data from Australia during 1983-1986

Quarter	Deaths	Quarter	Deaths	Quarter	Deaths
1	0	6	4	11	20
2	1	7	9	12	25
3	2	8	18	13	37
4	3	9	23	14	45
5	1	10	31		

Table 8.2 contains quarterly data on AIDS deaths in Australia that illustrate the application of a generalized linear model [32, 121]. A simple plot of the data suggests exponential growth. A plausible model therefore involves Poisson distributed observations y_i with means $\mu_i(\theta) = e^{\theta_1 + i\theta_2}$. Because this parameterization renders scoring equivalent to Newton's method, scoring gives the quick convergence noted in Table 8.3. ■

TABLE 8.3. Scoring Iterates for the AIDS Model

Iteration	Step Halves	θ_1	θ_2
1	0	0.0000	0.0000
2	3	-1.3077	0.4184
3	0	0.6456	0.2401
4	0	0.3744	0.2542
5	0	0.3400	0.2565
6	0	0.3396	0.2565

8.7 Problems

1. What happens when you apply Newton's method to the functions

$$f(x) = \begin{cases} \sqrt{x} & x \geq 0 \\ -\sqrt{-x} & x < 0 \end{cases}$$

and $g(x) = \sqrt[3]{x}$?

2. Consider a function $f(x) = (x-r)^k g(x)$ with a root r of multiplicity k. If $g'(x)$ is continuous at r, and the Newton iterates $x_{(m)}$ converge to r, then show that the iterates satisfy

$$\lim_{m \to \infty} \frac{|x_{m+1} - r|}{|x_m - r|} = 1 - \frac{1}{k}.$$

3. Suppose the real-valued $f(x)$ satisfies $f'(x) > 0$ and $f''(x) > 0$ for all x in its domain (d, ∞). If $f(x) = 0$ has a root, then demonstrate that the root is unique and that Newton's method converges to the root regardless of its starting point. When do the Newton iterates x_n converge monotonically to the root? Cite an example from the current chapter pertinent to this problem.

4. Prove that the increment (8.5) can be expressed as

$$\begin{aligned}
&\Delta x_{(m)} \\
&= -H_{(m)}^{-1/2}\left[I - H_{(m)}^{-1/2}V^*(VH_{(m)}^{-1}V^*)^{-1}VH_{(m)}^{-1/2}\right]H_{(m)}^{-1/2}\nabla f(x_{(m)}) \\
&= -H_{(m)}^{-1/2}(I - P_{(m)})H_{(m)}^{-1/2}\nabla f(x_{(m)})
\end{aligned}$$

using the symmetric square root $H_{(m)}^{-1/2}$ of $H_{(m)}^{-1}$. Check that the matrix $P_{(m)}$ is a projection in the sense that $P_{(m)}^* = P_{(m)}$ and $P_{(m)}^2 = P_{(m)}$ and that these properties carry over to $I - P_{(m)}$. Now argue that

$$-df(x_{(m)})\Delta x_{(m)} = \|(I - P_{(m)})H_{(m)}^{-1/2}\nabla f(x_{(m)})\|^2$$

and consequently that step halving is bound to produce a decrease in $f(x)$ if $(I - P_{(m)})H_{(m)}^{-1/2}\nabla f(x_{(m)}) \neq \mathbf{0}$.

5. In Example 8.4.2, digamma and trigamma functions $\psi(t)$ and $\psi'(t)$ must be evaluated. Show that these functions satisfy the recurrence relations

$$\begin{aligned}
\psi(t) &= -t^{-1} + \psi(t+1) \\
\psi'(t) &= t^{-2} + \psi'(t+1).
\end{aligned}$$

Thus, if $\psi(t)$ and $\psi'(t)$ can be accurately evaluated via asymptotic expansions for large t, then they can be accurately evaluated for small t. For example, it is known that $\psi(t) = \ln t - (2t)^{-1} + O(t^{-2})$ and $\psi'(t) = t^{-1} + (\sqrt{2}t)^{-2} + O(t^{-3})$ as $t \to \infty$.

6. Verify the formulas for the expected information matrix of the Poisson, exponential, and multinomial models mentioned in Section 8.5.

7. A quantal response model involves independent binomial observations $y_1, \ldots, y_j$ with n_i trials and success probability $\pi_i(\theta)$ per trial for the ith observation. If $x_{(i)}$ is a covariate vector and θ a parameter vector, then the specification

$$\pi_i(\theta) = \frac{e^{x_{(i)}^*\theta}}{1 + e^{x_{(i)}^*\theta}}$$

TABLE 8.4. Ingot Data for a Quantal Response Model

Trials n_i	Observation y_i	Covariate x_{i1}	Covariate x_{i2}
55	0	1	7
157	2	1	14
159	7	1	27
16	3	1	57

gives a generalized linear model. Use the scoring algorithm to estimate $\hat{\theta} = (-5.1316, 0.0677)^*$ for the ingot data of Cox [21] displayed in Table 8.4.

8. In robust regression it is useful to consider location-scale families with densities of the form

$$\frac{c}{\sigma} e^{-\rho(\frac{x-\mu}{\sigma})}, \qquad x \in (-\infty, \infty). \tag{8.15}$$

Here $\rho(r)$ is a strictly convex even function, decreasing to the left of 0 and symmetrically increasing to the right of 0. Without loss of generality, one can take $\rho(0) = 0$. The normalizing constant c is determined by $c \int_{-\infty}^{\infty} e^{-\rho(r)} dr = 1$. Show that a random variable X with density (8.15) has mean μ and variance

$$\mathrm{Var}(X) = c\sigma^2 \int_{-\infty}^{\infty} r^2 e^{-\rho(r)} dr.$$

If μ depends on a parameter vector θ, demonstrate that the score corresponding to a single observation $X = x$ amounts to

$$\nabla L(\phi) = \begin{pmatrix} \frac{1}{\sigma}\rho'(\frac{x-\mu}{\sigma})\nabla\mu(\theta) \\ -\frac{1}{\sigma} + \rho'(\frac{x-\mu}{\sigma})\frac{x-\mu}{\sigma^2} \end{pmatrix}$$

for $\phi = (\theta^*, \sigma)^*$. Finally, prove that the expected information $J(\phi)$ is block diagonal with upper-left block

$$\frac{c}{\sigma^2} \int_{-\infty}^{\infty} \rho''(r) e^{-\rho(r)} dr \nabla\mu(\theta) d\mu(\theta)$$

and lower-right block

$$\frac{c}{\sigma^2} \int_{-\infty}^{\infty} \rho''(r) r^2 e^{-\rho(r)} dr + \frac{1}{\sigma^2}.$$

9. In the context of Problem 8, take $\rho(r) = \ln \cosh^2(\frac{r}{2})$. Show that this corresponds to the logistic distribution with density

$$f(x) = \frac{e^{-x}}{(1 + e^{-x})^2}.$$

Compute the integrals

$$\begin{aligned} \frac{\pi^2}{3} &= c\int_{-\infty}^{\infty} r^2 e^{-\rho(r)}\,dr \\ \frac{1}{3} &= c\int_{-\infty}^{\infty} \rho''(r) e^{-\rho(r)}\,dr \\ \frac{1}{3}+\frac{\pi^2}{9} &= c\int_{-\infty}^{\infty} \rho''(r) r^2 e^{-\rho(r)}\,dr + 1 \end{aligned}$$

determining the variance and expected information of the density (8.15) for this choice of $\rho(r)$.

10. Continuing Problems 8 and 9, compute the normalizing constant c and the three integrals determining the variance and expected information for Huber's function

$$\rho(r) = \begin{cases} \frac{r^2}{2} & |r| \le k \\ k|r| - \frac{k^2}{2} & |r| > k\,. \end{cases}$$

11. A family of discrete density functions $p_n(\theta)$ defined on $\{0, 1, \ldots\}$ and indexed by a parameter $\theta > 0$ is said to be a power series family if for all n

$$p_n(\theta) = \frac{c_n \theta^n}{g(\theta)}, \tag{8.16}$$

where $c_n \ge 0$, and where $g(\theta) = \sum_{n=0}^{\infty} c_n \theta^n$ is the appropriate normalizing constant. If $y_1, \ldots, y_j$ are independent observations from the discrete density (8.16), then show that the maximum likelihood estimate of θ is a root of the equation

$$\frac{1}{j}\sum_{i=1}^{j} y_i = \frac{\theta g'(\theta)}{g(\theta)}.$$

Prove that the expected information in a single observation is

$$J(\theta) = \frac{\sigma^2(\theta)}{\theta^2},$$

where $\sigma^2(\theta)$ is the variance of the density (8.16).

12. In the Gauss-Newton algorithm (8.8), the matrix

$$d^2 f(\theta_{(m)}) = \sum_{i=1}^{j} w_i \nabla \mu_i(\theta_{(m)}) d\mu(\theta_{(m)})$$

can be singular or nearly so. To cure this ill, Marquardt suggested choosing $\lambda > 0$, substituting

$$H_{(m)} = \sum_{i=1}^{j} w_i \nabla \mu_i(\theta_{(m)}) d\mu(\theta_{(m)}) + \lambda I$$

for $d^2 f(\theta_{(m)})$, and iterating according to

$$\theta_{(m+1)} = \theta_{(m)} + H_{(m)}^{-1} \sum_{i=1}^{j} w_i [x_i - \mu_i(\theta_{(m)})] \nabla \mu_i(\theta_{(m)}). \quad (8.17)$$

Prove that the increment $\Delta\theta_{(m)} = \theta_{(m+1)} - \theta_{(m)}$ proposed in equation (8.17) minimizes the criterion

$$\frac{1}{2} \sum_{i=1}^{j} w_i [x_i - \mu_i(\theta_{(m)}) - d\mu(\theta_{(m)}) \Delta\theta_{(m)}]^2 + \frac{\lambda}{2} \|\Delta\theta_{(m)}\|^2.$$

13. Continuing Example 8.2.5, consider iterating according to

$$B_{(m+1)} = B_{(m)} \sum_{i=0}^{j} (I - AB_{(m)})^i \quad (8.18)$$

to find A^{-1} [67]. Example 8.2.5 is the special case $j = 1$. Verify the alternative representation

$$B_{(m+1)} = \sum_{i=0}^{j} (I - B_{(m)} A)^i B_{(m)},$$

and use it to prove that $B_{(m+1)}$ is symmetric whenever A and $B_{(m)}$ are. Also show by induction that

$$A^{-1} - B_{(m+1)} = (A^{-1} - B_{(m)})[A(A^{-1} - B_{(m)})]^j.$$

From this last identity deduce the norm inequality

$$\|A^{-1} - B_{(m+1)}\| \leq \|A\|^j \|A^{-1} - B_{(m)}\|^{j+1}.$$

Thus, the algorithm converges at a cubic rate when $j = 2$, at a quartic rate when $j = 3$, and so forth.

14. Example 8.2.2 can be adapted to extract the nth root of a positive semidefinite matrix A [49]. Consider the iteration scheme

$$B_{(m+1)} = \frac{n-1}{n} B_{(m)} + \frac{1}{n} B_{(m)}^{-n+1} A$$

starting with $B_{(0)} = cI$ for some positive constant c. Show by induction that (a) $B_{(m)}$ commutes with A, (b) $B_{(m)}$ is symmetric, and (c) $B_{(m)}$ is positive definite. To prove that $B_{(m)}$ converges to $A^{1/n}$, consider the spectral decomposition $A = UDU^*$ of A with D diagonal and U orthogonal. Show that $B_{(m)}$ has a similar spectral decomposition $B_{(m)} = UD_{(m)}U^*$ and that the ith diagonal entries of $D_{(m)}$ and D satisfy

$$d_{m+1,i} = \frac{n-1}{n} d_{mi} + \frac{1}{n} d_{mi}^{-n+1} d_i.$$

Example 8.2.2 implies that d_{mi} converges to $\sqrt[n]{d_i}$ when $d_i > 0$. This convergence occurs at a fast quadratic rate as explained in Proposition 10.2.2. If $d_i = 0$, then d_{mi} converges to 0 at the linear rate $\frac{n-1}{n}$.

15. Program the algorithm of Problem 14 and extract the square roots of the two matrices

$$\begin{pmatrix} 1 & 1 \\ 1 & 1 \end{pmatrix}, \quad \begin{pmatrix} 2 & 1 \\ 1 & 2 \end{pmatrix}.$$

Describe the apparent rate of convergence in each case and any difficulties you encounter with roundoff error.

9

Conjugate Gradient and Quasi-Newton

9.1 Introduction

Our discussion of Newton's method has highlighted both its strengths and its weaknesses. Related algorithms such as scoring and Gauss-Newton exploit special features of the objective function $f(x)$ in overcoming the defects of Newton's method. We now consider algorithms that apply to generic functions $f(x)$. These algorithms also operate by locally approximating $f(x)$ by a strictly convex quadratic function. Indeed, the guiding philosophy behind many modern optimization algorithms is to see what techniques work well with quadratic functions and then to modify the best techniques to generic functions.

The conjugate gradient algorithm [44, 60] is noteworthy for three properties: (a) it minimizes a quadratic function $f(x)$ from R^n to R in n steps, (b) it does not require evaluation of $d^2 f(x)$, and (c) it does not involve storage or inversion of any $n \times n$ matrices. Property (c) makes the method particularly suitable for optimization in high-dimensional settings. One of the drawbacks of the conjugate gradient method is that it requires exact line searches.

Quasi-Newton algorithms [5, 22, 41, 43] enjoy properties (a) and (b) but not property (c). In compensation for the failure of (c), inexact line searches are usually adequate with quasi-Newton algorithms. Furthermore, quasi-Newton methods adapt more readily to parameter constraints. Except for a discussion of trust regions, the current chapter considers only unconstrained optimization problems.

9.2 Centers of Spheres and Centers of Ellipsoids

As a gentle introduction to many of the central ideas of the chapter, it is instructive to explore a simple algorithm for finding the center of a sphere. The fact that we already know the answer should not deter us from considering algorithmic issues. If the center is the origin, then obviously we can find it by minimizing the scaled distance function

$$g(x) = \tfrac{1}{2}\|x\|^2 = \tfrac{1}{2}\textstyle\sum_{i=1}^n x_i^2$$

with gradient $\nabla g(x) = x$. In cyclic coordinate descent, we minimize $g(x)$ along each coordinate direction in turn, starting from a point $x_{(1)}$ and generating successive points $x_{(2)}, \ldots, x_{(n+1)}$. The search along coordinate direction $e_{(i)}$ at iteration i amounts to minimizing the function

$$g(x_{(i)} + te_{(i)}) = \frac{1}{2}(x_{ii} + t)^2 + \frac{1}{2}\sum_{j \neq i}^{n} x_{ij}^2$$

of the scalar t. The minimum occurs at $t = -x_{ii}$ and yields $x_{(i+1)}$. It is trivial to check that this procedure achieves the minimum in n iterations and satisfies at iteration i the identities

$$\begin{aligned} e_{(i)}^* e_{(j)} &= 0 \\ dg(x_{(i)})e_{(j)} &= 0 \end{aligned} \tag{9.1}$$

for all $j < i$. Furthermore, $x_{(i+1)}$ minimizes the function

$$h(t_1, \ldots, t_i) = g\Big(x_{(1)} + \sum_{j=1}^{i} t_j e_{(j)}\Big)$$

defined on the i-dimensional plane $x_{(1)} + t_1 e_{(1)} + \cdots + t_i e_{(i)}$ formed from all linear combinations of the first i search directions.

Because of the spherical symmetry of the function $g(x)$, we can substitute any set of nontrivial orthogonal vectors $u_{(1)}, \ldots, u_{(n)}$ and reach the same conclusions. If we consider an arbitrary strictly convex quadratic function

$$\begin{aligned} f(y) &= \frac{1}{2}y^* Ay + b^* y + c \qquad (9.2) \\ &= \frac{1}{2}(y + A^{-1}b)^* A(y + A^{-1}b) - \frac{1}{2}b^* A^{-1}b + c, \end{aligned}$$

then the situation becomes more interesting. Here the matrix A is positive definite, so there is no doubt that its inverse A^{-1} exists. We can reduce this ellipsoidal minimization problem to the previous spherical minimization problem by making the change of variables $y = A^{-1/2}x - A^{-1}b$ involving the symmetric square root $A^{-1/2}$ of A^{-1}. If $A = UDU^*$ is the spectral decomposition of A, then $A^{-1/2} = UD^{-1/2}U^*$. The invertible transformation $x \mapsto y$ sends lines into lines and planes into planes. It also sends the function $f(y)$ into the function

$$\begin{aligned} g(x) &= f(y) \\ &= f(A^{-1/2}x - A^{-1}b) \\ &= \frac{1}{2}\|x\|^2 - \frac{1}{2}b^* A^{-1} b + c \end{aligned}$$

and puts us back where we started, minimizing $\frac{1}{2}\|x\|^2$. If we have a set of nontrivial orthogonal vectors $u_{(1)}, \ldots, u_{(n)}$ in x space, then we can search along each of these directions in turn and achieve the global minima of $g(x)$ and $f(y)$ in n iterations.

For later use, it is important to identify the analogs of the orthogonality conditions (9.1). The direction $u_{(i)}$ in x space corresponds to the direction $v_{(i)}$ in y space defined by $A^{1/2}v_{(i)} = u_{(i)}$. Thus, the condition $u_{(i)}^* u_{(j)} = 0$ for all $j < i$ translates into the condition

$$v_{(i)}^* A^{1/2} A^{1/2} v_{(j)} = v_{(i)}^* A v_{(j)} = 0$$

for all $j < i$. Two vectors $v_{(i)}$ and $v_{(j)}$ satisfying such an orthogonality relation are said to be conjugate. Conjugacy is equivalent to orthogonality under the nonstandard inner product $v_{(i)}^* A v_{(j)}$.

In view of the chain rule, we have $dg(x_{(i)}) = df(y_{(i)})A^{-1/2}$ for the point $y_{(i)} = A^{-1/2}x_{(i)} - A^{-1}b$. Thus, the condition $dg(x_{(i)})u_{(j)} = 0$ for all $j < i$ translates into the condition

$$df(y_{(i)})A^{-1/2}A^{1/2}v_{(j)} = df(y_{(i)})v_{(j)} = 0 \tag{9.3}$$

for all $j < i$. Finally, the point $y_{(i+1)}$ minimizes the function

$$h(t_1, \ldots, t_i) = f\Big(y_{(1)} + \sum_{j=1}^{i} t_j v_{(j)}\Big)$$

defined on the i-dimensional plane $y_{(1)} + t_1 v_{(1)} + \cdots + t_i v_{(i)}$ formed from all linear combinations of the first i search directions.

9.3 The Conjugate Gradient Algorithm

The flaw with this analysis is that it omits any description of how the initial point $y_{(1)}$ and conjugate directions $v_{(1)}, \ldots, v_{(n)}$ are chosen. Choice of $y_{(1)}$ is more or less arbitrary, depending on the particular problem and relevant external information. The obvious choice $v_{(1)} = -\nabla f(y_{(1)})$ is consistent with an initial search along the direction of steepest descent. At iteration $i > 1$ the conjugate gradient algorithm inductively chooses the search direction

$$v_{(i)} = -\nabla f(y_{(i)}) + \alpha_i v_{(i-1)}, \tag{9.4}$$

where

$$\alpha_i = \frac{df(y_{(i)})Av_{(i-1)}}{v_{(i-1)}^* Av_{(i-1)}} \tag{9.5}$$

is defined so that $v_{(i)}^* Av_{(i-1)} = 0$. For $1 \le j < i-1$, the conjugacy condition $v_{(i)}^* Av_{(j)} = 0$ requires

$$0 = -df(y_{(i)})Av_{(j)} + \alpha_i v_{(i-1)}^* Av_{(j)} = -df(y_{(i)})Av_{(j)}. \tag{9.6}$$

Equality (9.6) is hardly obvious, but we can attack it by noting that the vectors $v_{(1)}, \ldots, v_{(i-1)}$ and $\nabla f(y_{(1)}), \ldots, \nabla f(y_{(i-1)})$ span the same vector subspace due to definition (9.4). Hence, the condition $df(y_{(i)})v_{(j)} = 0$, valid for all $j < i$, is equivalent to the condition $df(y_{(i)})\nabla f(y_{(j)}) = 0$ for all $j < i$. Because $\nabla f(y) = Ay + b$ and $y_{(j+1)} = y_{(j)} + t_j v_{(j)}$ for some optimal constant t_j, we can also write

$$\nabla f(y_{(j+1)}) = \nabla f(y_{(j)}) + t_j Av_{(j)}. \tag{9.7}$$

It follows that

$$\begin{aligned} df(y_{(i)})Av_{(j)} &= \frac{1}{t_j} df(y_{(i)})[\nabla f(y_{(j+1)}) - \nabla f(y_{(j)})] \\ &= 0 \end{aligned}$$

for $1 \le j < i-1$. Except for a few small details, this demonstrates equality (9.6) and completes the proof that the search directions $v_{(1)}, \ldots, v_{(n)}$ are conjugate.

If at any iteration we have $\nabla f(y_{(i)}) = \mathbf{0}$, then the algorithm terminates prematurely with the global minimum. Otherwise, equations (9.3) and (9.4) show that all search directions $v_{(i)}$ satisfy

$$\begin{aligned} df(y_{(i)})v_{(i)} &= -\|\nabla f(y_{(i)})\|^2 + \alpha_i df(y_{(i)})v_{(i-1)} \\ &= -\|\nabla f(y_{(i)})\|^2 \\ &< 0. \end{aligned}$$

As a consequence, $v_{(i)} \neq \mathbf{0}$ and α_{i+1} is well defined. Finally, the inequality $df(y_{(i)})v_{(i)} < 0$ implies that the search direction $v_{(i)}$ leads downhill from $y_{(i)}$ and that the search constant $t_i > 0$. In fact, using equation (9.7) one can easily demonstrate that

$$t_i = -\frac{df(y_{(i)})v_{(i)}}{v_{(i)}^* Av_{(i)}} = -\frac{(y_{(i)}^* A + b^*)v_{(i)}}{v_{(i)}^* Av_{(i)}}. \tag{9.8}$$

In generalizing the conjugate gradient algorithm to non-quadratic functions, we preserve most of the structure of the algorithm. Thus, the algorithm first searches along the negative gradient $v_{(1)} = -\nabla f(y_{(1)})$ emanating from the initial point $y_{(1)}$. At iteration i it searches along the direction $v_{(i)}$ defined by equality (9.4), avoiding the explicit formula (9.8) for t_i. The formula (9.5) for α_i is problematic because it appears to require A. Hestenes and Stiefel recommend the alternative formula

$$\alpha_i = \frac{df(y_{(i)})[\nabla f(y_{(i)}) - \nabla f(y_{(i-1)})]}{v_{(i-1)}[\nabla f(y_{(i)}) - \nabla f(y_{(i-1)})]} \tag{9.9}$$

based on the substitution

$$Av_{(i-1)} = \frac{1}{t_{i-1}}[\nabla f(y_{(i)}) - \nabla f(y_{(i-1)})].$$

Polak and Ribière suggest the further substitutions

$$\begin{aligned} v_{(i-1)}^* \nabla f(y_{(i)}) &= 0 \\ v_{(i-1)}^* \nabla f(y_{(i-1)}) &= [-df(y_{(i-1)}) + \alpha_{i-1} v_{(i-2)}] \nabla f(y_{(i-1)}) \\ &= -\|\nabla f(y_{(i-1)})\|^2. \end{aligned}$$

These produce the second alternative

$$\alpha_i = \frac{df(y_{(i)})[\nabla f(y_{(i)}) - \nabla f(y_{(i-1)})]}{\|\nabla f(y_{(i-1)})\|^2}. \tag{9.10}$$

Finally, Fletcher and Reeves note that the identity

$$\begin{aligned} df(y_{(i)}) \nabla f(y_{(i-1)}) &= df(y_{(i)})[-v_{(i-1)} + \alpha_{i-1} v_{(i-2)}] \\ &= 0 \end{aligned}$$

leads to the third alternative

$$\alpha_i = \frac{\|\nabla f(y_{(i)})\|^2}{\|\nabla f(y_{(i-1)})\|^2}. \tag{9.11}$$

Almost no one now uses the Hestenes-Stiefel update (9.9). Current opinion is divided between the Polak-Ribière update (9.10) and the Fletcher-Reeves update (9.11). *Numerical Recipes* [105] codes both formulas but leans toward the Polak-Ribière formula. In addition to this issue, there are other practical concerns in implementing the conjugate gradient algorithm. For example, if we fail to stop once the gradient $\nabla f(y)$ vanishes, then the Polak-Ribière and Fletcher-Reeves updates are undefined. This suggests stopping when $\|\nabla f(y_{(i)})\| \leq \epsilon \|\nabla f(y_{(0)})\|$ for some small $\epsilon > 0$. There is also the problem of loss of conjugacy. Assuming $f(y)$ is defined on R^n, it is common practice to restart the conjugate gradient algorithm with the steepest descent direction every n iterations. This is also a good idea whenever the descent condition $df(y_{(i)})v_{(i)} < 0$ fails. Finally, the algorithm is incomplete without specifying a line search algorithm. The next section discusses some of the possibilities. The references [2, 42, 105] provide a fuller account and computer code for line searches.

9.4 Line Search Methods

The secant method of Example 8.2.4 can obviously be adapted to minimize the objective function $f(y)$ along a line $r \mapsto x + rv$ in R^n. If we define $g(r) = f(x + rv)$, then we proceed by searching for a zero of the derivative $g'(r) = df(x+rv)v$. In this guise, the secant method is known as the method of false position. It iterates according to

$$r_{m+1} = r_m - \frac{g'(r_m)(r_{m-1} - r_m)}{g'(r_{m-1}) - g'(r_m)}.$$

Two criticisms of the method of false position immediately come to mind. One is that it indiscriminately heads for maxima as well as minima. Another is that it does not make full use of the available information.

A better alternative to the method of false position is to approximate $g(r)$ by a cubic polynomial matching the values of $g(r)$ and $g'(r)$ at r_m and r_{m-1}. Minimizing the cubic should lead to an improved estimate r_{m+1} of the minimum of $g(r)$. It simplifies matters notationally to rescale the interval by setting $h(s) = g(r_{m-1} + sd_m)$ and $d_m = r_m - r_{m-1}$. Now $s = 0$ corresponds to r_{m-1} and $s = 1$ corresponds to r_m. Furthermore, the chain rule implies $h'(s) = g'(r_{m-1} + sd_m)d_m$. Given these conventions, the theory of Hermite interpolation [58] suggests approximating $h(s)$ by the cubic polynomial

$$\begin{aligned}
& p(s) \\
= \;& (s-1)^2 h_0 + s^2 h_1 + s(s-1)[(s-1)(h_0' + 2h_0) + s(h_1' - 2h_1)] \\
= \;& (2h_0 + h_0' - 2h_1 + h_1')s^3 + (-3h_0 - 2h_0' + 3h_1 - h_1')s^2 + h_0's + h_0,
\end{aligned}$$

where $h_0 = h(0)$, $h_0' = h'(0)$, $h_1 = h(1)$, and $h_1' = h'(1)$. One can readily verify that $p(0) = h_0$, $p'(0) = h_0'$, $p(1) = h_1$, and $p'(1) = h_1'$.

The conjugate gradient method is locally descending in the sense that $p'(0) = h_0' < 0$. To be on the cautious side, $p'(1) = h_1' > 0$ should hold and $p(s)$ should be convex throughout the interval $[0, 1]$. To check convexity, it suffices to check the conditions $p''(0) \geq 0$ and $p''(1) \geq 0$ since $p''(s)$ is linear. Straightforward calculation shows that

$$\begin{aligned}
p''(0) &= -6h_0 + 6h_1 - 4h_0' - 2h_1' \\
p''(1) &= 6h_0 - 6h_1 + 2h_0' + 4h_1'.
\end{aligned}$$

Thus, $p(s)$ is convex throughout $[0,1]$ if and only if

$$\tfrac{1}{3}h_1' + \tfrac{2}{3}h_0' \;\leq\; h_1 - h_0 \;\leq\; \tfrac{2}{3}h_1' + \tfrac{1}{3}h_0'. \tag{9.12}$$

Under these conditions, a local minimum of $p(s)$ occurs on $[0,1]$. The pertinent root of the two possible roots of $p'(s) = 0$ is determined by the sign of the coefficient $2h_0 + h_0' - 2h_1 + h_1'$ of s^3 in $p(s)$. If this coefficient is positive, then the right root furnishes the minimum, and if this coefficient is negative, then the left root furnishes the minimum. The two roots can be calculated simultaneously by solving the quadratic equation

$$\begin{aligned} p'(s) &= 3(2h_0 + h_0' - 2h_1 + h_1')s^2 + 2(-3h_0 - 2h_0' + 3h_1 - h_1')s + h_0' \\ &= 0. \end{aligned}$$

If the condition $p'(1) = h_1' > 0$ or the convexity conditions (9.12) fail, or if the minimum of the cubic leads to an increase in $g(r)$, then one should fall back on more conservative search methods. Golden search involves recursively bracketing a minimum by three points $a < b < c$ satisfying $g(b) < \min\{g(a), g(c)\}$. The analogous method of bisection brackets a zero of $g(r)$ by two points $a < b$ satisfying $g(a)g(b) < 0$. For the moment we ignore the question of how the initial three points a, b, and c are chosen in golden search.

To replace the bracketing interval (a,c) by a shorter interval, we choose $d \in (a,c)$ so that d belongs to the longer of the two intervals (a,b) and (b,c). Without loss of generality, suppose $b < d < c$. If $g(d) < g(b)$, then the three points $b < d < c$ bracket a minimum. If $g(d) > g(b)$, then the three points $a < b < d$ bracket a minimum. In the case of a tie $g(d) = g(b)$, we choose $b < d < c$ when $g(c) < g(a)$ and $a < b < d$ when $g(a) < g(c)$.

These sensible rules do not address the problem of choosing d. Consider the fractional distances

$$\beta = \tfrac{b-a}{c-a}, \qquad \delta = \tfrac{d-b}{c-a}$$

along the interval (a,c). The next bracketing interval will have a fractional length of either $1-\beta$ or $\beta+\delta$. To guard against the worst case, we should take $1-\beta = \beta+\delta$. This determines $\delta = 1-2\beta$ and hence d. One could leave matters as they now stand, but the argument is taken one step further in Golden search. If we imagine repeatedly performing golden search, then scale similarity is expected to set in eventually so that

$$\beta = \frac{b-a}{c-a} = \frac{d-b}{c-b} = \frac{\delta}{1-\beta}.$$

Substituting $\delta = 1-2\beta$ in this identity and cross multiplying give the quadratic $\beta^2 - 3\beta + 1 = 0$ with solution

$$\beta = \frac{3-\sqrt{5}}{2}$$

equal to the golden mean of ancient Greek mathematics. Following this reasoning, we should take $d = \sqrt{5} - 2 = 0.2361$.

There is little theory to guide us in finding an initial bracketing triple $a < b < c$. It is clear that $a = 0$ is one natural choice. In view of the condition $g'(0) < 0$, the point b can be chosen close to 0 as well. This leaves c, which is usually selected based on specific knowledge of $f(y)$, parabolic extrapolation, or repeated doubling of some small arbitrary distance.

9.5 Quasi-Newton Methods

Quasi-Newton methods of minimization update the current approximation $H_{(i)}$ to the second differential $d^2 f(x_{(i)})$ of the objective function $f(x)$ by a low-rank perturbation satisfying a secant condition. The secant condition originates from the first-order Taylor approximation

$$\nabla f(x_{(i+1)}) - \nabla f(x_{(i)}) \quad \approx \quad d^2 f(x_{(i)})(x_{(i+1)} - x_{(i)}).$$

If we set

$$\begin{aligned} g_{(i)} &= \nabla f(x_{(i+1)}) - \nabla f(x_{(i)}) \\ d_{(i)} &= x_{(i+1)} - x_{(i)}, \end{aligned}$$

then the secant condition reads $H_{(i+1)} d_{(i)} = g_{(i)}$. The unique, symmetric, rank-one update to $H_{(i)}$ satisfying the secant condition is furnished by Davidon's formula [22]

$$H_{(i+1)} = H_{(i)} + c_i v_{(i)} v_{(i)}^* \tag{9.13}$$

with constant c_i and vector $v_{(i)}$ specified by

$$\begin{aligned} c_i &= -\frac{1}{(H_{(i)} d_{(i)} - g_{(i)})^* d_{(i)}} \qquad (9.14) \\ v_{(i)} &= H_{(i)} d_{(i)} - g_{(i)}. \end{aligned}$$

An immediate concern is that the constant c_i is undefined when the inner product $(H_{(i)} d_{(i)} - g_{(i)})^* d_{(i)} = 0$. In such situations or when

$$|(H_{(i)} d_{(i)} - g_{(i)})^* d_{(i)}| \ll \|H_{(i)} d_{(i)} - g_{(i)}\| \cdot \|d_{(i)}\|,$$

then the secant adjustment is ignored, and the value $H_{(i)}$ is retained for $H_{(i+1)}$.

We have stressed the desirability of maintaining a positive definite approximation $H_{(i)}$ to the second differential $d^2 f(x_{(i)})$. Because this is not always possible with the rank-one update, numerical analysts have investigated rank-two updates. The involvement of the vectors $g_{(i)}$ and $H_{(i)} d_{(i)}$ in the rank-one update suggests trying a rank-two update of the form

$$H_{(i+1)} = H_{(i)} + b_i g_{(i)} g_{(i)}^* + c_i H_{(i)} d_{(i)} d_{(i)}^* H_{(i)}. \tag{9.15}$$

Taking the product of both sides of this equation with $d_{(i)}$ gives

$$H_{(i+1)}d_{(i)} = H_{(i)}d_{(i)} + b_i g_{(i)} g_{(i)}^* d_{(i)} + c_i H_{(i)} d_{(i)} d_{(i)}^* H_{(i)} d_{(i)}.$$

To achieve consistency with the secant condition $H_{(i+1)}d_{(i)} = g_{(i)}$, we set

$$b_i = \frac{1}{g_{(i)}^* d_{(i)}}, \qquad c_i = -\frac{1}{d_{(i)}^* H_{(i)} d_{(i)}}.$$

The resulting rank-two update was proposed by Broyden, Fletcher, Goldfarb, and Shanno and is consequently known as the BFGS update.

The symmetric rank-one update (9.13) certainly preserves positive definiteness when $c_i \geq 0$. If $c_i < 0$, then $H_{(i+1)}$ is positive definite only if

$$\begin{aligned} v_{(i)}^* H_{(i)}^{-1}[H_{(i)} + c_i v_{(i)} v_{(i)}^*] H_{(i)}^{-1} v_{(i)} &= v_{(i)}^* H_{(i)}^{-1} v_{(i)} [1 + c_i v_{(i)}^* H_{(i)}^{-1} v_{(i)}] \\ &> 0. \end{aligned}$$

In other words,

$$1 + c_i v_{(i)}^* H_{(i)}^{-1} v_{(i)} > 0 \tag{9.16}$$

must hold. Conversely, condition (9.16) is sufficient to guarantee positive definiteness of $H_{(i+1)}$. This fact can be most easily demonstrated by noting the Sherman-Morrison inversion formula [97]

$$\begin{aligned} & [H_{(i)} + c_i v_{(i)} v_{(i)}^*]^{-1} \\ = \ & H_{(i)}^{-1} - \frac{c_i}{1 + c_i v_{(i)}^* H_{(i)}^{-1} v_{(i)}} H_{(i)}^{-1} v_{(i)} [H_{(i)}^{-1} v_{(i)}]^*. \end{aligned} \tag{9.17}$$

Formula (9.17) shows that $[H_{(i)} + c_i v_{(i)} v_{(i)}^*]^{-1}$ exists and is positive definite under condition (9.16). Since the inverse of a positive definite matrix is positive definite, it follows that $H_{(i)} + c_i v_{(i)} v_{(i)}^*$ is positive definite as well.

If $c_i < 0$ in the rank-one update, then it may be necessary to shrink c_i to maintain positive definiteness. Unfortunately, condition (9.16) gives too little guidance. Problem 7 shows how to control the size of $\det H_{(i+1)}$ while simultaneously forcing positive definiteness. An even better strategy that monitors the condition number of $H_{(i+1)}$ rather than $\det H_{(i+1)}$ is sketched in Problem 9. Finally, there is the option of using c_i as defined but perturbing $H_{(i+1)}$ by adding a constant multiple μI of the identity matrix. This tactic is closely related to the trust region method discussed in Section 9.6. If λ_1 is the smallest eigenvalue of $H_{(i+1)}$, then $H_{(i+1)} + \mu I$ is positive definite whenever $\lambda_1 + \mu > 0$. Problem 10 discusses a fast algorithm for finding λ_1. With an appropriate safeguard of positive definiteness in place, some numerical analysts [20, 75] consider the rank-one update superior to the BFGS update.

Positive definiteness is almost automatic with the BFGS update (9.15). The key turns out to be the inequality

$$0 \ < \ g_{(i)}^* d_{(i)} \ = \ df(x_{(i+1)})d_{(i)} - df(x_{(i)})d_{(i)}. \qquad (9.18)$$

This is ordinarily true for two reasons. First, because $d_{(i)}$ is proportional to the current search direction $v_{(i)} = -H_{(i)}^{-1}\nabla f(x_{(i)})$, positive definiteness of $H_{(i)}^{-1}$ implies $-df(x_{(i)})d_{(i)} > 0$. Second, when a full search is conducted, the identity $df(x_{(i+1)})d_{(i)} = 0$ holds. Even a partial search typically entails condition (9.18). Section 10.4 takes up the issue of partial line searches. The speed of partial line searches compared to that of full line searches makes quasi-Newton methods superior to the conjugate gradient method on small-scale problems.

To show that the BFGS update $H_{(i+1)}$ is positive definite when condition (9.18) holds, we examine the quadratic form

$$u^* H_{(i+1)} u \ = \ u^* H_{(i)} u + \frac{[g_{(i)}^* u]^2}{g_{(i)}^* d_{(i)}} - \frac{[u^* H_{(i)} d_{(i)}]^2}{d_{(i)}^* H_{(i)} d_{(i)}} \qquad (9.19)$$

for $u \neq \mathbf{0}$. The Cauchy-Schwarz inequality applied to the vectors $a = H_{(i)}^{1/2} u$ and $b = H_{(i)}^{1/2} d_{(i)}$ gives

$$[u^* H_{(i)} d_{(i)}]^2 \ \leq \ [u^* H_{(i)} u][d_{(i)}^* H_{(i)} d_{(i)}],$$

with equality if and only if u is proportional to $d_{(i)}$. Hence, the sum of the first and third terms on the right of equality (9.19) is nonnegative. In the event that u is proportional to $d_{(i)}$, the second term on the right of equality (9.19) is positive by assumption. It follows that $u^* H_{(i+1)} u > 0$ and therefore that $H_{(i+1)}$ is positive definite.

In successful applications of quasi-Newton methods, choice of the initial matrix $H_{(1)}$ is critical. Setting $H_{(1)} = I$ is convenient but often poorly scaled for a particular problem. In maximum likelihood estimation, the expected information matrix $J(x_{(1)})$, if available, is preferable to the identity matrix. In some problems, $J(x)$ is cheap to compute and manipulate for special values of x. For instance, $J(x)$ may be diagonal in certain circumstances. These special x should be considered as starting points for a quasi-Newton search.

It is possible to carry forward approximations $K_{(i)}$ of $d^2 f(x_{(i)})^{-1}$ rather than of $d^2 f(x_{(i)})$. This tactic has the advantage of avoiding matrix inversion in computing the quasi-Newton search direction $v_{(i)} = -K_{(i)}\nabla f(x_{(i)})$. The basic idea is to restate the secant condition $H_{(i+1)} d_{(i)} = g_{(i)}$ as the inverse secant condition $K_{(i+1)} g_{(i)} = d_{(i)}$. This substitution leads to the symmetric rank-one update

$$K_{(i+1)} \ = \ K_{(i)} + c_i w_{(i)} w_{(i)}^*, \qquad (9.20)$$

where $c_i = -[(K_{(i)}g_{(i)} - d_{(i)})^* g_{(i)}]^{-1}$ and $w_{(i)} = K_{(i)}g_{(i)} - d_{(i)}$. Note that monitoring positive definiteness of $K_{(i)}$ is still an issue.

For a rank-two update, our earlier arguments apply provided we interchange the roles of $d_{(i)}$ and $g_{(i)}$. The Davidon-Fletcher-Powell (DFP) update

$$K_{(i+1)} = K_{(i)} + b_i d_{(i)} d_{(i)}^* + c_i K_{(i)} g_{(i)} g_{(i)}^* K_{(i)} \qquad (9.21)$$

with

$$b_i = \frac{1}{g_{(i)}^* d_{(i)}}, \qquad c_i = -\frac{1}{g_{(i)}^* K_{(i)} g_{(i)}}$$

is a competitor to the BFGS update, but the consensus seems to be that the BFGS update is superior to the DFP update in numerical practice [26].

In closing this section, we would like to prove that the BFGS algorithm with an exact line search converges in n or fewer iterations for the strictly convex quadratic function (9.2) defined on R^n. Recall that at iteration i we search along the direction $v_{(i)} = -H_{(i)}^{-1}\nabla f(x_{(i)})$ and then in preparation for the next iteration construct $H_{(i+1)}$ according to the BFGS formula (9.15). Unless $\nabla f(x_{(i)}) = \mathbf{0}$ and the iterates converge prematurely, the step increment $d_{(i)}$ is a positive multiple $t_i v_{(i)}$ of the search direction $v_{(i)}$. Our proof of convergence consists of a subtle inductive argument proving two claims in parallel. These claims amount to the equalities $v_{(i+1)} A v_{(j)} = 0$ and $H_{(i+1)} d_{(j)} = g_{(j)}$ for all $i \le n$ and all $j \le i$. Given the efficacy of successive searches along conjugate directions as demonstrated in Section 9.2, the conjugacy condition $v_{(i+1)} A v_{(j)} = 0$ guarantees convergence to the minimum of $f(y)$ in n or fewer iterations.

The identity $H_{(i+1)} d_{(i)} = g_{(i)}$ is just the secant requirement imposed in defining $H_{(i+1)}$. The case $i = 1$ gets the induction on i started for the extended secant claim. Assuming that $H_{(i+1)} d_{(j)} = g_{(j)}$ and that the current gradient $\nabla f(x_{(i+1)})$ is perpendicular to the previous search directions $v_{(1)}, \ldots, v_{(i)}$, we calculate

$$\begin{aligned} v_{(i+1)}^* A v_{(j)} &= -df(x_{(i+1)}) H_{(i+1)}^{-1} A v_{(j)} \\ &= -t_j^{-1} df(x_{(i+1)}) H_{(i+1)}^{-1} A d_{(j)} \\ &= -t_j^{-1} df(x_{(i+1)}) H_{(i+1)}^{-1} g_{(j)} \\ &= -t_j^{-1} df(x_{(i+1)}) d_{(j)} \\ &= -df(x_{(i+1)}) v_{(j)} \\ &= 0. \end{aligned} \qquad (9.22)$$

The first of these assumptions obviously holds for $i = j = 1$. The second is a special case of the perpendicularity condition noted in equation (9.3) for conjugate directions $v_{(1)}, \ldots, v_{(i)}$. Thus, both claims hold for $i = 1$.

Now assume that both claims are true for $i-1$ and all $j \leq i-1$. Revisiting the calculation (9.22) advances the conjugacy claim provided both the extended secant claim and the perpendicularity condition are true. Once again equation (9.3) validates the perpendicularity condition.

Thus, it suffices to prove $H_{(i+1)}d_{(j)} = g_{(j)}$ for $j \leq i$. Again the case $j = i$ is just the secant requirement. For $j < i$, we observe that

$$H_{(i+1)}d_{(j)} = H_{(i)}d_{(j)} + b_i g_{(i)} g_{(i)}^* d_{(j)} + c_i H_{(i)} d_{(i)} d_{(i)}^* H_{(i)} d_{(j)}.$$

Now the equalities $H_{(i)}d_{(j)} = g_{(j)}$ and $g_{(i)}^* d_{(j)} = d_{(i)}^* A d_{(j)} = 0$ hold by the induction hypothesis. Likewise,

$$\begin{aligned} d_{(i)}^* H_{(i)} d_{(j)} &= d_{(i)}^* g_{(j)} \\ &= d_{(i)}^* A d_{(j)} \\ &= 0 \end{aligned}$$

follows from the induction hypothesis. Combining these equalities makes it clear that $H_{(i+1)}d_{(j)} = g_{(j)}$ and completes the induction and the proof.

9.6 Trust Regions

If the quadratic approximation

$$f(x) \approx f(x_{(i)}) + df(x_{(i)})(x - x_{(i)}) + \frac{1}{2}(x - x_{(i)})^* H_{(i)}(x - x_{(i)})$$

to the objective function $f(x)$ is poor, then the naive step

$$x_{(i+1)} = x_{(i)} - H_{(i)}^{-1} \nabla f(x_{(i)}) \tag{9.23}$$

computed in a quasi-Newton method may be absurdly large. This situation often occurs for early iterates. One remedy is to minimize the quadratic approximation to $f(x)$ subject to the spherical constraint $\frac{1}{2}\|x - x_{(i)}\|^2 \leq \frac{1}{2}r^2$ for a fixed radius r. This constrained optimization problem has a solution regardless of whether $H_{(i)}$ is positive definite. To guarantee a unique solution, we assume in the remaining discussion that $H_{(i)}$ is positive definite. According to Proposition 4.2.1, the solution satisfies the multiplier rule

$$\mathbf{0} = \nabla f(x_{(i)}) + H_{(i)}(x - x_{(i)}) + \mu(x - x_{(i)}) \tag{9.24}$$

for some nonnegative constant μ. If the minimum point x occurs within the open ball $\|x - x_{(i)}\| < r$, then $\mu = 0$ and x coincides with the naive step $x_{(i+1)}$ of equation (9.23). In this case the naive step is reasonable if it leads to a decrease in $f(x)$. If it does not, then one can always step halve or conduct an inexact search for a better point along the line segment between $x_{(i)}$ and $x_{(i+1)}$.

If the minimum point x satisfies $\|x - x_{(i)}\| = r$, then the multiplier μ may be positive. The fact that μ is unknown makes it impossible to find the minimum in closed form. In principle, one can overcome this difficulty by solving for μ iteratively, say by Newton's method. Hence, we view equation (9.24) as defining $x - x_{(i)}$ as a function of μ and ask for the value of μ that yields $\|x - x_{(i)}\| = r$. To simplify notation, let $M = H_{(i)}$, $y = x - x_{(i)}$, and $e = -\nabla f(x_{(i)})$. We now seek a zero of the function

$$\phi(\mu) = \frac{1}{r} - \frac{1}{\|y(\mu)\|} \qquad (9.25)$$

with $y(\mu)$ defined by $(M + \mu I)y(\mu) = e$. Note that $\phi(0) > 0$ if and only if $\|y(0)\| > r$. To implement Newton's method, we need $\phi'(\mu)$. An easy calculation shows that

$$\phi'(\mu) = \frac{y(\mu)^* y'(\mu)}{\|y(\mu)\|^3}.$$

Unfortunately, this formula contains the unknown derivative $y'(\mu)$. However, differentiation of the equation $(M + \mu I)y(\mu) = e$ readily yields

$$y(\mu) + (M + \mu I)y'(\mu) = \mathbf{0}, \qquad (9.26)$$

which implies $y'(\mu) = -(M + \mu I)^{-1}y(\mu)$. The complete formula

$$\phi'(\mu) = -\frac{y(\mu)^*(M + \mu I)^{-1}y(\mu)}{\|y(\mu)\|^3}$$

shows that $\phi(\mu)$ is strictly decreasing. Problem 11 asks the reader to calculate $\phi''(\mu)$ and verify that it is nonnegative. Problem 12 asserts that $\phi(\mu)$ is negative for large μ. Hence, there is a unique Lagrange multiplier $\mu_i > 0$ solving $\phi(\mu) = 0$ whenever $\phi(0) > 0$. The corresponding $y(\mu_i)$ solves the trust region problem.

This elegant solution was discovered by Moré and Sorensen [98] and elaborated by Sorensen [112]. For problems with many parameters, there are quicker methods such as the "hookstep" and the "dogleg" for dealing with trust regions [26, 74].

9.7 Problems

1. Suppose that you possess n conjugate vectors $v_{(1)}, \ldots, v_{(n)}$ for the positive definite matrix A. Describe how you can use these to solve the linear equation $Ax = b$.

2. Suppose that A is an $n \times n$ positive definite matrix and that the nontrivial vectors $u_{(1)}, \ldots, u_{(n)}$ satisfy

$$u_{(i)}^* A u_{(j)} = 0$$
$$u_{(i)}^* u_{(j)} = 0$$

for all $i \neq j$. Demonstrate that the $u_{(i)}$ are eigenvectors of A.

3. Consider the quadratic function

$$Q(x) = \frac{1}{2}x^* \begin{pmatrix} 2 & 1 \\ 1 & 1 \end{pmatrix} x + (1,1)x$$

defined on R^2. Compute by hand the iterates of the conjugate gradient and BFGS algorithms starting from $x_{(1)} = (0,0)^*$. For the BFGS algorithm take $H_{(1)} = I$ and use an exact line search. You should find that the two sequences of iterates coincide. This phenomenon holds more generally for any strictly convex quadratic function given $H_{(1)} = I$ in the BFGS algorithm [100].

4. Write a Fortran, C, or MATLAB program to implement the conjugate gradient algorithm. Apply it to the function

$$f(x) = \frac{1}{4}x_1^4 + \frac{1}{2}x_2^2 - x_1x_2 + x_1 - x_2$$

with two local minima. Demonstrate that your program will converge to either minimum depending on its starting value.

5. Prove Woodbury's generalization

$$(A + UBV^*)^{-1} = A^{-1} - A^{-1}U(B^{-1} + V^*A^{-1}U)^{-1}V^*A^{-1}$$

of the Sherman-Morrison matrix inversion formula for compatible matrices A, B, U, and V. Apply this formula to the BFGS rank-two update.

6. A quasi-Newton minimization of the strictly convex quadratic function (9.2) generates a sequence of points $x_{(1)}, \ldots, x_{(n+1)}$ with A-conjugate differences $d_{(i)} = x_{(i+1)} - x_{(i)}$. At the final iteration we have argued that the approximate Hessian satisfies $H_{(n+1)}d_{(i)} = g_{(i)}$ for $1 \leq i \leq n$ and $g_{(i)} = \nabla f(x_{(i+1)}) - \nabla f(x_{(i)})$. Show that this implies $H_{(n+1)} = A$.

7. In the rank-one update, suppose we want both $H_{(i+1)}$ to remain positive definite and $\det H_{(i+1)}$ to exceed some constant $\epsilon > 0$. Explain how these criteria can be simultaneously met by replacing $c_i < 0$ by

$$\max\left\{c_i, \left(\frac{\epsilon}{\det H_{(i)}} - 1\right)\frac{1}{v_{(i)}^* H_{(i)}^{-1} v_{(i)}}\right\}$$

in updating $H_{(i)}$. (Hint: In verifying this sufficient condition, you may want to use the one-dimensional version of the identity

$$\det(A)\det(B^{-1} - U^*A^{-1}U) = \det(A - UBU^*)\det(B^{-1})$$

for compatible matrices A, B, and U.)

8. Let H be a positive definite matrix. Prove [14] that

$$\mathrm{tr}(H) - \ln\det(H) \quad \geq \quad \ln[\mathrm{cond}_2(H)]. \qquad (9.27)$$

The condition number $\mathrm{cond}_2(H)$ of H equals $\|H\| \cdot \|H^{-1}\|$, that is the ratio of the largest to smallest eigenvalue of H. (Hint: Express $\mathrm{tr}(H) - \ln\det(H)$ in terms of the eigenvalues of H. Then use the inequalities $\lambda - \ln\lambda \geq 1$ and $\lambda > 2\ln\lambda$ for all $\lambda > 0$.)

9. In Davidon's symmetric rank-one update (9.13), it is possible to control the condition number of $H_{(i+1)}$ by shrinking the constant c_i. Suppose a moderately sized number δ is chosen. Due to inequality (9.27), one can avoid ill-conditioning in the matrices $H_{(i)}$ by imposing the constraint $\mathrm{tr}(H_{(i)}) - \ln\det(H_{(i)}) \leq \delta$. To see how this fits into the updating scheme (9.13), verify that

$$\begin{aligned}\ln\det(H_{(i+1)}) &= \ln\det(H_{(i)}) + \ln(1 + c_i v_{(i)}^* H_{(i)}^{-1} v_{(i)}) \\ \mathrm{tr}(H_{(i+1)}) &= \mathrm{tr}(H_{(i)}) + c_i\|v_{(i)}\|_2^2.\end{aligned}$$

Employing these results, deduce that $\mathrm{tr}(H_{(i+1)}) - \ln\det(H_{(i+1)}) \leq \delta$ provided c_i satisfies

$$c_i\|v_{(i)}\|_2^2 - \ln(1 + c_i v_{(i)}^* H_{(i)}^{-1} v_{(i)}) \quad \leq \quad \delta - \mathrm{tr}(H_{(i)}) + \ln\det(H_{(i)}).$$

10. Suppose the $n \times n$ symmetric matrix A has eigenvalues

$$\lambda_1 < \lambda_2 \leq \cdots \leq \lambda_{n-1} < \lambda_n.$$

The iterative scheme $x_{(i+1)} = (A - \eta_i I)x_{(i)}$ can be used to approximate either λ_1 or λ_n. Consider the criterion

$$\sigma_i \quad = \quad \frac{x_{(i+1)}^* A x_{(i+1)}}{x_{(i+1)}^* x_{(i+1)}}.$$

Choosing η_i to maximize σ_i causes $\lim_{i\to\infty}\sigma_i = \lambda_n$, while choosing η_i to minimize σ_i causes $\lim_{i\to\infty}\sigma_i = \lambda_1$. Show that the extrema of σ_i as a function of η are given by the roots of the quadratic equation

$$0 \quad = \quad \det\begin{pmatrix} 1 & \eta & \eta^2 \\ \tau_0 & \tau_1 & \tau_2 \\ \tau_1 & \tau_2 & \tau_3 \end{pmatrix},$$

where $\tau_k = x_{(i)}^* A^k x_{(i)}$. Apply this algorithm to find the largest and smallest eigenvalue of the matrix

$$A \quad = \quad \begin{pmatrix} 10 & 7 & 8 & 7 \\ 7 & 5 & 6 & 5 \\ 8 & 6 & 10 & 9 \\ 7 & 5 & 9 & 10 \end{pmatrix}.$$

You should find $\lambda_1 = 0.01015$ and $\lambda_4 = 30.2887$ [18].

11. Calculate the second derivative $\phi''(\mu)$ of the function defined in equation (9.25). Prove that $\phi''(\mu) \geq 0$.

12. Show that the solution $y(\mu)$ of the equation $(M+\mu I)y(\mu) = e$ satisfies $\lim_{\mu \to \infty} \|y(\mu)\| = 0$. How does this justify the conclusion that the function (9.25) has a zero on $(0, \infty)$ when $\phi(0) > 0$?

10
Analysis of Convergence

10.1 Introduction

Proving convergence of the various optimization algorithms is a delicate exercise. In general, it is helpful to consider local and global convergence patterns separately. The local convergence rate of an algorithm provides a useful benchmark for comparing it to other algorithms. On this basis, Newton's method wins hands down. However, the tradeoffs are subtle. Besides the sheer number of iterations until convergence, the computational complexity and numerical stability of an algorithm are critically important. The MM algorithm is often the epitome of numerical stability and computational simplicity. Scoring lies somewhere between Newton's method and the MM algorithm. It tends to converge more quickly than the MM algorithm and to behave more stably than Newton's method. Quasi-Newton methods also occupy this intermediate zone. Because the issues are complex, all of these algorithms survive and prosper in certain computational niches.

The following short overview of convergence manages to cover only highlights. For the sake of simplicity, only unconstrained problems are treated. Quasi-Newton methods are also ignored. The efforts of a generation of numerical analysts in understanding quasi-Newton methods defy easy summary or digestion. Interested readers can consult one of the helpful references [26, 48, 89, 102]. We emphasize MM and gradient algorithms, partially because a fairly coherent theory for them can be reviewed in a few pages.

10.2 Local Convergence

Local convergence of many optimization algorithms hinges on the following result [103].

Proposition 10.2.1 (Ostrowski) *Let the differentiable map $h(x)$ from an open set $U \subset \mathsf{R}^n$ into R^n have fixed point y. If $\|dh(y)\|_\dagger < 1$ for some induced matrix norm, and if $x_{(0)}$ is sufficiently close to y, then the iterates $x_{(m+1)} = h(x_{(m)})$ converge to y.*

Proof: Let $h(x)$ have slope function $s(x, y)$ near y. For any constant r satisfying $\|dh(y)\|_\dagger < r < 1$, we have $\|s(x, y)\| \leq r$ for x sufficiently close to y. It therefore follows from the identities

$$x_{(m+1)} - y \quad = \quad h(x_{(m)}) - h(y) \quad = \quad s(x_{(m)}, y)(x_{(m)} - y)$$

that a proper choice of $x_{(0)}$ yields

$$\|x_{(m+1)} - y\|_\dagger \quad \leq \quad \|s(x_{(m)}, y)\|_\dagger \|x_{(m)} - y\|_\dagger \quad \leq \quad r\|x_{(m)} - y\|_\dagger.$$

In other words, the distance from $x_{(m)}$ to y contracts by a factor of at least r at every iteration. This proves convergence. ■

Two comments are worth making about Proposition 10.2.1. First, the appearance of a general vector norm and its induced matrix norm obscures the fact that the condition $\rho[dh(y)] < 1$ on the spectral radius of $dh(y)$ is the operative criterion. One can prove that any induced matrix norm exceeds the spectral radius and that some induced matrix norm comes within ϵ of it for any small $\epsilon > 0$ [82]. Later in this section, we will generate a tight matrix norm by taking an $n \times n$ invertible matrix T and forming $\|u\|_T = \|Tu\|$. It is easy to check that this defines a legitimate vector norm and by setting $v = Tu$ that the induced matrix norm $\|M\|_T$ on $n \times n$ matrices M satisfies

$$\|M\|_T \quad = \quad \sup_{u \neq \mathbf{0}} \frac{\|TMu\|}{\|Tu\|} \quad = \quad \sup_{v \neq \mathbf{0}} \frac{\|TMT^{-1}v\|}{\|v\|}.$$

In other words, $\|M\|_T = \|TMT^{-1}\|$, and we are back in the familiar terrain covered by the Euclidean norm.

Our second comment involves two definitions. A sequence $x_{(m)}$ is said to converge linearly to a point y at rate $r < 1$ provided

$$\lim_{m \to \infty} \frac{\|x_{(m+1)} - y\|}{\|x_{(m)} - y\|} \quad = \quad r.$$

The sequence converges quadratically if the limit

$$\lim_{m \to \infty} \frac{\|x_{(m+1)} - y\|}{\|x_{(m)} - y\|^2} \quad = \quad c$$

exists. Ostrowski's result guarantees at least linear convergence; Newton's method improves linear convergence to quadratic convergence.

Our intention is to apply Ostrowski's result to iteration maps of the type

$$h(x) \quad = \quad x - A(x)^{-1}b(x). \tag{10.1}$$

In optimization problems, $b(x) = \nabla f(x)$ for some real-valued function $f(x)$ defined on R^n. The matrix $A(x)$ is then typically $d^2 f(x)$ or a positive definite or negative definite approximation to it. For instance, in statistical applications, $-A(x)$ could be either the observed or expected information. In the MM gradient algorithm, $A(x)$ is the second differential $d^2 g(x \mid x_{(m)})$ of the surrogate function. Our first order of business is to compute the differential $dh(y)$ and an associated slope function $s_h(x,y)$ at a fixed point y of $h(x)$ in terms of the slope function $s_b(x,y)$ of $b(x)$. Because $b(y) = \mathbf{0}$ at a fixed point, the calculation

$$\begin{aligned} h(x) - h(y) \quad &= \quad x - y - A(x)^{-1}[b(x) - b(y)] \\ &= \quad [I - A(x)^{-1}s_b(x,y)](x - y) \end{aligned} \tag{10.2}$$

identifies the possibly asymmetric slope function

$$s_h(x,y) \quad = \quad I - A(x)^{-1}s_b(x,y)$$

and corresponding differential

$$dh(y) \quad = \quad I - A(y)^{-1}db(y). \tag{10.3}$$

There is no harm in the asymmetry here.

In Newton's method, $A(y)^{-1} = db(y)$ and

$$I - A(y)^{-1}db(y) \quad = \quad I - db(y)^{-1}db(y) \quad = \quad \mathbf{0}.$$

Proposition 10.2.1 therefore implies that the Newton iterates are locally attracted to a fixed point y. Of course, this conclusion is predicated on the suppositions that $db(x)$ tends to $db(y)$ as x tends to y and that $db(y)$ is invertible. To demonstrate quadratic rather than linear convergence, we now assume that $b(x)$ possesses a second differential at y. If $s_b^2(x,y)$ denotes the second slope function of $b(x)$, then

$$\begin{aligned} I - db(x)^{-1}s_b(x,y) \quad &= \quad db(x)^{-1}[db(x) - s_b(x,y)] \\ &= \quad \frac{1}{2}db(x)^{-1}I \otimes (x-y)^* s_b^2(x,y), \end{aligned}$$

and consequently

$$\|I - db(x)^{-1}s_b(x,y)\| \;\leq\; \frac{1}{2}\|db(x)^{-1}\| \cdot \|I \otimes (x-y)^*\| \cdot \|s_b^2(x,y)\|. \tag{10.4}$$

To make further progress, we must find the norm of the Kronecker product $I \otimes (x-y)^*$. Fortunately, this is easy because the norm of a generic Kronecker product $C \otimes D$ satisfies

$$\|C \otimes D\| = \|C\| \cdot \|D\|.$$

To validate this claim, let $C^*C = U\Sigma U^*$ and $D^*D = V\Omega V^*$ be the standard diagonalizations of C^*C and D^*D. Then

$$\begin{aligned}(C \otimes D)^*(C \otimes D) &= (C^* \otimes D^*)(C \otimes D) \\ &= (C^*C) \otimes (D^*D) \\ &= (U\Sigma U^*) \otimes (V\Omega V^*) \\ &= (U \otimes V)(\Sigma \otimes \Omega)(U^* \otimes V^*) \\ &= (U \otimes V)(\Sigma \otimes \Omega)(U \otimes V)^*.\end{aligned}$$

Each eigenvalue of $(C\otimes D)^*(C\otimes D)$ appears on the diagonal of the diagonal matrix $\Sigma \otimes \Omega$ as a product of a diagonal entry σ_i of Σ times a diagonal entry ω_j of Ω. The result now follows from the identity

$$\max_{ij} |\sigma_i \omega_j| = \max_i |\sigma_i| \max_j |\omega_j|.$$

To demonstrate the quadratic convergence of Newton's method, it is just a matter of assembling the various pieces. From equality (10.2), inequality (10.4), and the formula $\|I \otimes (x-y)^*\| = \|x-y\|$, we deduce the quadratic bound

$$\begin{aligned}\|x_{(m+1)} - y\| &\leq \|I - db(x)^{-1}s_b(x,y)\| \cdot \|x_{(m)} - y\| \\ &\leq \frac{1}{2}\|db(x)^{-1}\| \cdot \|s_b^2(x,y)\| \cdot \|x_{(m)} - y\|^2.\end{aligned}$$

Our results can summarized as follows:

Proposition 10.2.2 *Let y be a fixed point of the function $b(x)$ from an open set $U \subset \mathsf{R}^n$ into R^n. Suppose that $b(x)$ is differentiable in a neighborhood of y and twice differentiable at y, If $db(y)$ is invertible, then Newton's method converges to y at a quadratic rate or better whenever $x_{(0)}$ is sufficiently near y.*

Proof: The preceding remarks make it clear that

$$\limsup_{m\to\infty} \frac{\|x_{(m+1)} - y\|}{\|x_{(m)} - y\|^2} \leq \frac{1}{2}\|db(y)^{-1}\| \cdot \|d^2b(y)\|,$$

and this is enough for at least quadratic convergence. ■

We now turn to the MM gradient algorithm. Suppose we are minimizing $f(x)$ via the surrogate function $g(x \mid x_{(m)})$. If y is a local minimum of $f(x)$, it is reasonable to assume that the matrices $C = d^2f(y)$ and $D = d^2g(y \mid y)$ are positive definite. Because $g(x \mid y) - f(x)$ attains its minimum at $x = y$, the matrix difference $D - C$ is certainly positive semidefinite. The MM gradient algorithm iterates with $A(x) = d^2g(x \mid x)$. In view of formula (10.3), the iteration map $h(x)$ has differential $I - D^{-1}C$ at y. If we let T be the symmetric square root $D^{1/2}$ of D, then

$$\begin{aligned} I - D^{-1}C &= D^{-1}(D - C) \\ &= T^{-1}T^{-1}(D - C)T^{-1}T. \end{aligned}$$

Hence, $I - D^{-1}C$ is similar to $T^{-1}(D - C)T^{-1}$.

To establish local attraction of the MM gradient algorithm to y, we need to choose an appropriate induced matrix norm. A good choice is $\|\cdot\|_T$, for in this norm Example 1.4.3 and Proposition 2.2.1 imply that

$$\begin{aligned} \|I - D^{-1}C\|_T &= \|TT^{-1}T^{-1}(D - C)T^{-1}TT^{-1}\| \\ &= \|T^{-1}(D - C)T^{-1}\| \\ &= \sup_{u \neq \mathbf{0}} \frac{u^*T^{-1}(D - C)T^{-1}u}{u^*u} \\ &= \sup_{v \neq \mathbf{0}} \frac{v^*(D - C)v}{v^*T^*Tv} \\ &= \sup_{v \neq \mathbf{0}} \frac{v^*(D - C)v}{v^*Dv} \\ &= 1 - \inf_{\|v\|=1} \frac{v^*Cv}{v^*Dv}. \end{aligned}$$

The symmetry and positive semidefiniteness of $D - C$ come into play in the third equality in this string of equalities. By virtue of the positive definiteness of C and D, the continuous ratio v^*Cv/v^*Dv is bounded below by a positive constant on the compact sphere $\{v : \|v\| = 1\}$. It follows that $\|I - D^{-1}C\|_T < 1$, and Ostrowski's result applies. Hence, the iterates $x_{(m)}$ are locally attracted to y.

Calculation of the differential $dh(y)$ of an MM iteration map $h(x)$ is equally interesting. The next iterate solves the equation

$$\nabla g(x_{(m+1)} \mid x_{(m)}) = \mathbf{0}.$$

Assuming that $d^2g(y \mid y)$ is invertible, the implicit function theorem, Proposition 3.8.2, shows that the iteration map $h(x)$ satisfies

$$\nabla g[h(x) \mid x] = \mathbf{0}$$

and is continuously differentiable with differential

$$dh(x) = -d^2g[h(x) \mid x]^{-1}d^{11}g[h(x) \mid x]. \tag{10.5}$$

Here $d^{11}g(u \mid v)$ denotes the differential of $dg(u \mid v)$ with respect to v. At the fixed point y of $h(x)$, equation (10.5) becomes

$$dh(y) = -d^2g(y \mid y)^{-1}d^{11}g(y \mid y). \tag{10.6}$$

Further simplification can be achieved by taking the differential of

$$df(x) - dg(x \mid x) = \mathbf{0}^*$$

and setting $x = y$. These actions give

$$d^2f(y) - d^2g(y \mid y) - d^{11}g(y \mid y) = \mathbf{0}.$$

This last equation can be solved for $d^{11}g(y \mid y)$, and the result substituted in equation (10.6). It follows that

$$\begin{aligned} dh(y) &= -d^2g(y \mid y)^{-1}[d^2f(y) - d^2g(y \mid y)] \\ &= I - d^2g(y \mid y)^{-1}d^2f(y), \end{aligned} \tag{10.7}$$

which is precisely the differential computed for the MM gradient algorithm. Hence, the MM and MM gradient algorithms display exactly the same behavior in converging to a stationary point of $f(x)$.

Local convergence of the scoring algorithm is not guaranteed by Proposition 10.2.1 because nothing prevents an eigenvalue of

$$dh(y) = I + J(y)^{-1}d^2L(y)$$

from falling below -1. Here $L(x)$ is the loglikelihood, $J(x)$ is the expected information, and $h(x)$ is the scoring iteration map. Scoring with a fixed partial step,

$$x_{(m+1)} = x_{(m)} + tJ(x_{(m)})^{-1}\nabla L(x_{(m)}),$$

will converge locally for $t > 0$ sufficiently small. In practice, no adjustment is usually necessary. For reasonably large sample sizes, the expected information matrix $J(y)$ approximates the observed information matrix $-d^2L(y)$ well, and the spectral radius of $dh(y)$ is nearly 0.

10.3 Global Convergence of the MM Algorithm

In this section and the next, we tackle global convergence. We begin with the MM algorithm and consider without loss of generality minimization of the objective function $f(x)$ via the majorizing surrogate $g(x \mid x_{(m)})$. In studying global convergence, we must carefully specify the parameter domain U. Let us take U to be any open convex subset of R^n. To avoid colliding with the boundary of U, we assume that $f(x)$ is coercive in the sense that the set $\{x \in U : f(x) \leq f(z)\}$ is compact for every $z \in U$. Coerciveness implies that $f(x)$ attains its minimum somewhere in U and that $f(x)$ tends to ∞ as x approaches the boundary of U or as $\|x\|$ approaches ∞. It is convenient to continue assuming when necessary that $f(x)$ and $g(x \mid x_{(m)})$ and their various first and second differentials are jointly continuous in x and $x_{(m)}$.

We also demand that the second differential $d^2g(x \mid x_{(m)})$ be positive definite. This implies that $g(x \mid x_{(m)})$ is strictly convex. Strict convexity in turn implies that the solution $x_{(m+1)}$ of the minimization step is unique. Existence of a solution fortunately is guaranteed by coerciveness. Indeed, the closed set

$$\{x : g(x \mid x_{(m)}) \leq g(x_{(m)} \mid x_{(m)}) = f(x_{(m)})\}$$

is compact because it is contained within the compact set

$$\{x : f(x) \leq f(x_{(m)})\}.$$

Finally, the implicit function theorem, Proposition 3.8.2, shows that the iteration map $x_{(m+1)} = M(x_{(m)})$ is continuously differentiable in a neighborhood of every point $x_{(m)}$. Local differentiability of $M(x)$ clearly extends to global differentiability.

Gradient versions of the algorithm (10.1) have the property that stationary points of the objective function and fixed points of the iteration map coincide. This property also applies to the MM algorithm. Here we recall the two identities $dg(x_{(m+1)} \mid x_{(m)}) = \mathbf{0}^*$ and $dg(x_{(m)} \mid x_{(m)}) = df(x_{(m)})$ and the strict convexity of $g(x \mid x_{(m)})$. By the same token, stationary points and only stationary points give equality in the descent inequality $f[M(x)] \leq f(x)$.

The next technical result prepares the ground for a proof of global convergence. We remind the reader that a point y is a limit point of a sequence $x_{(m)}$ provided there is a subsequence $x_{(m_k)}$ that tends to y. One can easily verify that any limit of a sequence of limit points is also a limit point and that a bounded sequence has a limit if and only if it has at most one limit point. See Problem 14.

Proposition 10.3.1 *If a bounded sequence $x_{(m)}$ in R^n satisfies*

$$\lim_{m \to \infty} \|x_{(m+1)} - x_{(m)}\| = 0, \tag{10.8}$$

then its set T of limit points is connected. If T is finite, then T reduces to a single point, and $\lim_{m \to \infty} x_{(m)} = y$ exists.

Proof: It is straightforward to prove that T is a compact set. If it is disconnected, then there is a continuous disconnecting function $\phi(x)$ having exactly the two values 0 and 1. The inverse images of the closed sets 0 and 1 under $\phi(x)$ can be represented as the intersections $T_0 = T \cap C_0$ and

$T_1 = T \cap C_1$ of T with two closed sets C_0 and C_1. Because T is compact, T_0 and T_1 are closed, nonempty, and disjoint. Furthermore, the distance

$$d(T_0, T_1) = \inf_{u \in T_0} d(u, T_1) = \inf_{u \in T_0, v \in T_1} \|u - v\|$$

separating T_0 and T_1 is positive. Indeed, the continuous function $d(u, T_1)$ attains its minimum at some point u of the compact set T_0, and the distance $d(u, T_1)$ separating that u from T_1 must be positive because T_1 is closed.

Now consider the sequence $x_{(m)}$ in the statement of the proposition. For large enough m, we have $\|x_{(m+1)} - x_{(m)}\| < d(T_0, T_1)/4$. As the sequence $x_{(m)}$ bounces back and forth between limit points in T_0 and T_1, it must enter the closed set $W = \{u : d(u, T) \geq d(T_0, T_1)/4\}$ infinitely often. But this means that W contains a limit point of $x_{(m)}$. Because W is disjoint from T_0 and T_1, and these two sets are postulated to contain all of the limit points of $x_{(m)}$, this contradiction implies that T is connected.

Because a finite set with more than one point is necessarily disconnected, T can be a finite set only if it consists of a single point. Finally, a bounded sequence with only a single limit point has that point as its limit. ■

With these facts in mind, we now state and prove a version of Liapunov's theorem for discrete dynamical systems [89].

Proposition 10.3.2 (Liapunov) *Let Γ be the set of limit points generated by the sequence $x_{(m+1)} = M(x_{(m)})$ starting from some initial $x_{(0)}$. Then Γ is contained in the set S of stationary points of $f(x)$.*

Proof: The sequence $x_{(m)}$ stays within the compact set

$$\{x \in U : f(x) \leq f(x_{(0)})\}.$$

Consider a limit point $z = \lim_{k\to\infty} x_{(m_k)}$. Since the sequence $f(x_{(m)})$ is monotone decreasing and bounded below, $\lim_{m\to\infty} f(x_{(m)})$ exists. Hence, taking limits in the inequality $f[M(x_{(m_k)})] \leq f(x_{(m_k)})$ and using the continuity of $M(x)$ and $f(x)$, we infer that $f[M(z)] = f(z)$. Thus, z is a fixed point of $M(x)$ and consequently also a stationary point of $f(x)$. ■

The next two propositions are adapted from [96]. In the second of these, recall that a point x in a set S is isolated if and only if there exists a radius $r > 0$ such that $S \cap B(x, r) = \{x\}$.

Proposition 10.3.3 *The set of limit points Γ of $x_{(m+1)} = M(x_{(m)})$ is compact and connected.*

Proof: Γ is a closed subset of the compact set $\{x \in U : f(x) \leq f(x_{(0)})\}$ and is therefore itself compact. According to Proposition 10.2.1, Γ is connected provided $\lim_{m\to\infty} \|x_{(m+1)} - x_{(m)}\| = 0$. If this sufficient condition fails, then the compactness of $\{x \in U : f(x) \leq f(x_{(0)})\}$ makes it possible to extract a subsequence $x_{(m_k)}$ such that $\lim_{k\to\infty} x_{(m_k)} = u$ and $\lim_{k\to\infty} x_{(m_k+1)} = v$

both exist, but $v \neq u$. However, the continuity of $M(x)$ requires $v = M(u)$ while the descent condition implies

$$f(v) = f(u) = \lim_{m \to \infty} f(x_{(m)}).$$

The equality $f(v) = f(u)$ forces the contradictory conclusion that u is a fixed point of $M(x)$. Hence, the sufficient condition (10.8) for connectivity holds. ■

Proposition 10.3.4 *Suppose that all stationary points of $f(x)$ are isolated and that the differentiability, coerciveness, and convexity assumptions are true. Then any sequence of iterates $x_{(m+1)} = M(x_{(m)})$ generated by the iteration map $M(x)$ of the MM algorithm possesses a limit, and that limit is a stationary point of $f(x)$. If $f(x)$ is strictly convex, then* $\lim_{m \to \infty} x_{(m)}$ *is the minimum point.*

Proof: In the compact set $\{x \in U : f(x) \leq f(x_{(0)})\}$ there can only be a finite number of stationary points. An infinite number of stationary points would admit a convergent sequence whose limit would not be isolated. Since the set of limit points Γ is a connected subset of this finite set of stationary points, Γ reduces to a single point. ■

Two remarks on Proposition 10.3.4 are in order. First, except when full strict convexity prevails for $f(x)$, the proposition offers no guarantee that the limit y of the sequence $x_{(m)}$ furnishes a global minimum. Problem 8 contains a counterexample of Wu [122] exhibiting convergence to a saddle point in the EM algorithm. Fortunately, in practice, descent algorithms almost always converge to at least a local minimum of the objective function. Second, if the set S of stationary points is not discrete, then there exists a sequence $z_{(m)} \in S$ converging to $z \in S$ with $z_{(m)} \neq z$ for all m. Because the surface of the unit sphere in R^n is compact, we can extract a subsequence such that

$$\lim_{k \to \infty} \frac{z_{(m_k)} - z}{\|z_{(m_k)} - z\|} = v$$

exists and is nontrivial. Taking limits in

$$\begin{aligned} \mathbf{0}^* &= \frac{1}{\|z_{(m_k)} - z\|}[df(z_{(m_k)}) - df(z)] \\ &= \frac{1}{2}\frac{1}{\|z_{(m_k)} - z\|}(z_{(m_k)} - z)^*\left[s_f^2(z_{(m_k)}, z) + s_f^2(z, z_{(m_k)})\right] \end{aligned}$$

then produces $\mathbf{0}^* = v^* d^2 f(z)$. In other words, the second differential at z is singular. If one can rule out such degeneracies, then all stationary points are isolated. Interested readers can consult the literature on Morse functions for further commentary on this subject [54].

10.4 Global Convergence of Gradient Algorithms

We now turn to the question of global convergence for gradient algorithms of the sort

$$x_{(m+1)} = x_{(m)} - A(x_{(m)})^{-1}\nabla f(x_{(m)}).$$

The assumptions concerning $f(x)$ made in the previous section remain in force. A major impediment to establishing the global convergence of any minimization algorithm is the possible failure of the descent property

$$f(x_{(m+1)}) \leq f(x_{(m)})$$

enjoyed by the MM algorithm. Provided the matrix $A(x_{(m)})$ is positive definite, the direction $v_{(m)} = -A(x_{(m)})^{-1}\nabla f(x_{(m)})$ is guaranteed to point locally downhill. Hence, if we elect the natural strategy of instituting a limited line search along the direction $v_{(m)}$ emanating from $x_{(m)}$, then we can certainly find an $x_{(m+1)}$ that decreases $f(x)$.

Although an exact line search is tempting, we may pay too great a price for precision when we need mere progress. The step-halving tactic mentioned in Chapter 8 is better than a full line search but not quite adequate for theoretical purposes. Instead, we require a sufficient decrease along a descent direction v. This is summarized by the Armijo rule of considering only steps tv satisfying the inequality

$$f(x+tv) \leq f(x) + \alpha t df(x)v \tag{10.9}$$

for t and some fixed α in $(0,1)$. To avoid too stringent a test, we take a low value of α such as 10^{-4}. In combining Armijo's rule with regular step decrementing, we first test the step v. If it satisfies Armijo's rule we are done. If it fails, we choose $\sigma \in (0,1)$ and test σv. It this fails, we test $\sigma^2 v$, and so forth until we encounter and take the first partial step $\sigma^k v$ that works. In step halving, obviously $\sigma = 1/2$.

Step halving can be combined with a partial line search. For instance, suppose the line search has been confined to the interval $t \in [0, s]$. If the point $x + sv$ passes Armijo's test, then we accept it. Otherwise, we fit a cubic to the function $t \mapsto f(x+tv)$ on the interval $[0, s]$ as described in Section 9.4. If the minimum point t of the cubic approximation satisfies $t \geq \sigma s$ and passes Armijo's test, then we accept $x + tv$. Otherwise, we replace the interval $[0, s]$ by the interval $[0, \sigma s]$ and proceed inductively. For the sake of simplicity in the sequel, we will ignore this elaboration of step halving and concentrate on the unadorned version.

We would like some guarantee that the exponent k of the step decrementing power σ^k does not grow too large. Mindful of this criterion, we suppose that the positive definite matrix $A(x)$ depends continuously on x. This is not much of a restriction for Newton's method, the Gauss-Newton algorithm, the MM gradient algorithm, or scoring. If we combine continuity with coerciveness, then we can conclude that there exist positive constants β, γ, and δ with

$$\|A(x)\| \;\le\; \beta, \quad \|A(x)^{-1}\| \;\le\; \gamma, \quad \|s_f^2(y,x)\| \;\le\; \delta$$

for all x and y in the compact set $\{x \in U : f(x) \le f(x_{(0)})\}$ where any descent algorithm acts. Here $s_f^2(y,x)$ is the second slope of $f(x)$.

To use these bounds, let $v = -A(x)^{-1}\nabla f(x)$ and consider the inequality

$$\begin{aligned} f(x+tv) &= f(x) + t df(x)v + \frac{1}{2}t^2 v^* s_f^2(x+tv,x)v \\ &\le f(x) + t df(x)v + \frac{1}{2}t^2\delta\|v\|^2. \end{aligned} \tag{10.10}$$

Taking into account the bound on $\|A(x)\|$ and the identity

$$\|A(x)^{1/2}\| \;=\; \|A(x)\|^{1/2}$$

entailed by Proposition 2.2.1, we also have

$$\begin{aligned} \|\nabla f(x)\|^2 &= \|A(x)^{1/2}A(x)^{-1/2}\nabla f(x)\|^2 \\ &\le \|A(x)^{1/2}\|^2\|A(x)^{-1/2}\nabla f(x)\|^2 \\ &\le \beta df(x)A(x)^{-1}\nabla f(x). \end{aligned} \tag{10.11}$$

It follows that

$$\begin{aligned} \|v\|^2 &= \|A(x)^{-1}\nabla f(x)\|^2 \\ &\le \gamma^2\|\nabla f(x)\|^2 \\ &= -\beta\gamma^2 df(x)v. \end{aligned}$$

Combining this last inequality with inequality (10.10) yields

$$f(x+tv) \;\le\; f(x) + t\left(1 - \frac{\beta\gamma^2\delta}{2}t\right) df(x)v.$$

Hence, as soon as σ^k satisfies

$$1 - \frac{\beta\gamma^2\delta}{2}\sigma^k \;\ge\; \alpha,$$

Armijo's rule (10.9) holds. In terms of k, backtracking is guaranteed to succeed in at most

$$k_{\max} \;=\; \left\lceil \frac{1}{\ln\sigma}\ln\frac{2(1-\alpha)}{\beta\gamma^2\delta}\right\rceil$$

decrements. Of course, a lower value of k may suffice.

Proposition 10.4.1 *Suppose that all stationary points of $f(x)$ are isolated and that the continuity, differentiability, positive definiteness, and coerciveness assumptions are true. Then any sequence of iterates generated by the iteration map $M(x) = x - tA(x)^{-1}\nabla f(x)$ with t chosen by step decrementing possesses a limit, and that limit is a stationary point of $f(x)$. If $f(x)$ is strictly convex, then $\lim_{m\to\infty} x_{(m)}$ is the minimum point.*

Proof: Let $v_{(m)} = -A(x_{(m)})^{-1}\nabla f(x_{(m)})$ and $x_{(m+1)} = x_{(m)} + \sigma^{k_m} v_{(m)}$. The sequence $f(x_{(m)})$ is decreasing by construction. Because the function $f(x)$ is bounded below on the compact set $\{x \in U : f(x) \leq f(x_{(0)})\}$, $f(x_{(m)})$ is bounded below as well and possesses a limit. Based on Armijo's rule (10.9) and inequality (10.11), we calculate

$$\begin{aligned} f(x_{(m)}) - f(x_{(m+1)}) &\geq -\alpha\sigma^{k_m} df(x_{(m)})v_{(m)} \\ &= \alpha\sigma^{k_m} df(x_{(m)})A(x_{(m)})^{-1}\nabla f(x_{(m)}) \\ &\geq \frac{\alpha\sigma^{k_m}}{\beta}\|\nabla f(x_{(m)})\|^2. \end{aligned}$$

Since $\sigma^{k_m} \geq \sigma^{k_{\max}}$, and the difference $f(x_{(m)}) - f(x_{(m+1)})$ tends to 0, we deduce that $\|\nabla f(x_{(m)})\|$ tends to 0. This conclusion and the inequality

$$\begin{aligned} \|x_{(m+1)} - x_{(m)}\| &= \sigma^{k_m}\|A(x_{(m)})^{-1}\nabla f(x_{(m)})\| \\ &\leq \sigma^{k_m}\gamma\|\nabla f(x_{(m)})\|, \end{aligned}$$

demonstrate that $\|x_{(m+1)} - x_{(m)}\|$ has limit 0 as well.

Given these results, Propositions 10.3.2 and 10.3.3 are true. All claims of the current proposition now follow as in the proof of Proposition 10.3.4. ■

10.5 Problems

1. Consider the functions $f(x) = x - x^3$ and $g(x) = x + x^3$. Show that the iterates $x_{m+1} = f(x_m)$ are locally attracted to 0 and that the iterates $x_{m+1} = g(x_m)$ are locally repelled by 0. In both cases $f'(0) = g'(0) = 1$.

2. Consider the iteration map $h(x) = \sqrt{a+x}$ on $(0, \infty)$ for $a > 0$. Find the fixed point of $h(x)$ and show that it is locally attractive. Is it also globally attractive?

3. Let A and B be $n \times n$ matrices. If A and $A - B^*AB$ are both positive definite, then show that B has spectral radius $\rho(B) < 1$. Note that A is symmetric, but B need not be symmetric.

4. In Example 8.2.1 suppose $x_0 = 1$ and $a \in (0, 2)$. Demonstrate that

$$\begin{aligned} x_m &= \frac{1-(1-a)^{2^m}}{a} \\ \left|x_{m+1} - \frac{1}{a}\right| &= a\left|x_m - \frac{1}{a}\right|^2. \end{aligned}$$

This shows very explicitly that x_m converges to $1/a$ at a quadratic rate.

5. In Example 8.2.2 prove that

$$\begin{aligned} x_m &= \sqrt{a} + \frac{2\sqrt{a}}{\left[\left(1+\frac{2\sqrt{a}}{x_0-\sqrt{a}}\right)^{2^m} - 1\right]} \\ \left|x_{m+1} - \sqrt{a}\right| &\leq \frac{1}{2\sqrt{a}}\left|x_m - \sqrt{a}\right|^2 \end{aligned}$$

when $n = 2$ and $x_0 > 0$. Thus, Newton's method converges at a quadratic rate. Use the first of these formulas or the iteration equation directly to show that $\lim_{m\to\infty} x_m = -\sqrt{a}$ for $x_0 < 0$.

6. Consider a Poisson distributed random variable Y with mean $a\theta + b$, where a and b are known positive constants and $\theta \geq 0$ is a parameter to be estimated. Devise an MM algorithm for maximum likelihood estimation of θ. If $Y = y$ is observed, then show that the MM iterates are

$$\theta_{m+1} = \frac{y\theta_m}{a\theta_m + b}.$$

Show that these iterates converge monotonically to the maximum likelihood estimate $\max\{0, (y-b)/a\}$. When $y = b$, verify that convergence to the boundary value 0 occurs at a rate slower than linear [40]. (Hint: When $y = b$ check that $\theta_m = b\theta_0/(ma\theta_0 + b)$.)

7. Consider independent observations $y_1, \ldots, y_n$ from the univariate t-distribution. These data have loglikelihood

$$\begin{aligned} L &= -\frac{n}{2}\ln\sigma^2 - \frac{\nu+1}{2}\sum_{i=1}^n \ln(\nu + \delta_i^2) \\ \delta_i^2 &= \frac{(y_i-\mu)^2}{\sigma^2}. \end{aligned}$$

To illustrate the occasionally bizarre behavior of the MM algorithm, we take $\nu = 0.05$ and the data vector $y = (-20, 1, 2, 3)^*$ with $n = 4$

observations. Devise an MM maximum likelihood algorithm for estimating μ with σ^2 fixed at 1. Show that the iteration map is

$$\begin{aligned}\mu_{m+1} &= \frac{\sum_{i=1}^n w_{mi} y_i}{\sum_{i=1}^n w_{mi}} \\ w_{mi} &= \frac{\nu+1}{\nu+(y_i-\mu_m)^2}.\end{aligned}$$

Plot the likelihood curve and show that it has the four local maxima -19.993, 1.086, 1.997, and 2.906 and the three local minima -14.516, 1.373, and 2.647. Demonstrate numerically convergence to a local maximum that is not the global maximum. Show that the algorithm converges to a local minimum in one step starting from -1.874 or -0.330 [93].

TABLE 10.1. Bivariate Normal Data for the EM Algorithm

Obs	Obs	Obs	Obs	Obs	Obs
(1,1)	(1,-1)	(-1,1)	(-1,-1)	(2,∗)	(2,∗)
(-2,∗)	(-2,∗)	(∗,2)	(∗,2)	(∗,-2)	(∗,-2)

8. Suppose the data displayed in Table 10.1 constitute a random sample from a bivariate normal distribution with both means 0, variances σ_1^2 and σ_2^2, and correlation coefficient ρ. The asterisks indicate missing values. Specify the EM algorithm for estimating σ_1^2, σ_2^2, and ρ. Show that the observed loglikelihood has global maxima at $\rho = \pm\frac{1}{2}$ and $\sigma_1^2 = \sigma_2^2 = \frac{8}{3}$ and a saddle point at $\rho = 0$ and $\sigma_1^2 = \sigma_2^2 = \frac{5}{2}$. If the EM algorithm starts with $\rho = 0$, prove that convergence to the saddle point occurs [122].

9. Under the hypotheses of Proposition 10.3.4, if the MM gradient algorithm is started close enough to a local minimum y of $f(x)$, then the iterates $x_{(m)}$ converge to y without step decrementing. Prove that for all sufficiently large m, either $x_{(m)} = y$ or $f(x_{(m+1)}) < f(x_{(m)})$ [81]. (Hints: Let $v_{(m)} = x_{(m+1)} - x_{(m)}$, $C_{(m)} = s_f^2(x_{(m+1)}, x_{(m)})$, and $D_{(m)} = d^2 g(x_{(m)} \mid x_{(m)})$. Show that

$$f(x_{(m+1)}) = f(x_{(m)} + \frac{1}{2} v_{(m)}^* [C_{(m)} - 2D_{(m)}] v_{(m)}.$$

Then use a continuity argument, noting that $d^2 g(y \mid y) - d^2 f(y)$ is positive semidefinite and $d^2 g(y \mid y)$ is positive definite.)

10. Let $M(x)$ be the MM algorithm or MM gradient algorithm map. Consider the modified algorithm $M_t(x) = x + t[M(x) - x]$ for $t > 0$. At

a local optimum y, show that the spectral radius ρ_t of the differential $dM_t(y) = (1-t)I + tdM(y)$ satisfies $\rho_t < 1$ when $0 < t < 2$. Hence, Ostrowski's theorem implies local attraction of $M_t(x)$ to y. If the largest and smallest eigenvalues of $dM(y)$ are $\omega_{\max}$ and $\omega_{\min}$, then prove that ρ_t is minimized by taking $t = [1 - (\omega_{\min} + \omega_{\max})/2]^{-1}$. In practice, the eigenvalues of $dM(y)$ are impossible to predict without advance knowledge of y, but for many problems the value $t = 2$ works well [81]. (Hints: All eigenvalues of $dM(y)$ occur on $[0, 1)$. To every eigenvalue ω of $dM(y)$, there corresponds an eigenvalue $\omega_t = 1-t+t\omega$ of $dM_t(y)$ and vice versa.)

11. Which of the following functions is coercive on its domain?

    (a) $f(x) = x + 1/x$ on $(0, \infty)$,
    (b) $f(x) = x - \ln x$ on $(0, \infty)$,
    (c) $f(x) = x_1^2 + x_2^2 - 2x_1x_2$ on R^2,
    (d) $f(x) = x_1^4 + x_2^4 - 3x_1x_2$ on R^2,
    (e) $f(x) = x_1^2 + x_2^2 + x_3^2 - \sin(x_1x_2x_3)$ on R^3.

    Give convincing reasons in each case.

12. Consider a polynomial $p(x)$ in n variables $x_1, \ldots, x_n$. Suppose that $p(x) = \sum_{i=1}^n c_i x_i^{2m}$ + lower-order terms, where all $c_i > 0$ and where a lower-order term is a product $bx_1^{m_1} \cdots x_n^{m_n}$ with $\sum_{i=1}^n m_i < 2m$. Prove rigorously that $p(x)$ is coercive on R^n.

13. Demonstrate that $h(x) + k(x)$ is coercive on R^n if $k(x)$ is convex and $h(x)$ satisfies $\lim_{\|x\|\to\infty} \|x\|^{-1}h(x) = \infty$. Problem 13 of Chapter 5 is a special case. (Hint: Apply definition (5.3) of Chapter 5.)

14. Consider a sequence $x_{(m)}$ in R^n. Verify that the set of limit points of $x_{(m)}$ is closed. If $x_{(m)}$ is bounded, then show that it has a limit if and only if it has at most one limit point.

15. In our exposition of least absolute deviation regression, we considered in Problem 9 of Chapter 6 a modified iteration scheme that minimizes the criterion

$$h_\epsilon(\theta) = \sum_{i=1}^p \left\{[y_i - \mu_i(\theta)]^2 + \epsilon\right\}^{1/2}. \qquad (10.12)$$

For a sequence of constants ϵ_m tending to 0, let $\theta_{(m)}$ be a corresponding sequence minimizing (10.12). If ϕ is a limit point of this sequence and the regression functions $\mu_i(\theta)$ are continuous, then show that ϕ minimizes $h_0(\theta) = \sum_{i=1}^p |y_i - \mu_i(\theta)|$. If, in addition, the minimum point ϕ of $h_0(\theta)$ is unique and $\lim_{\|\theta\|\to\infty} \sum_{i=1}^p |\mu_i(\theta)| = \infty$, then prove that $\lim_{m\to\infty} \theta_{(m)} = \phi$. (Hints: For the first assertion, take limits in

$$h_\epsilon(\theta_{(m)}) \leq h_\epsilon(\theta).$$

For the second assertion, it suffices to prove that the sequence $\theta_{(m)}$ is confined to a bounded set. To prove this fact, demonstrate the inequalities $|\mu| + |y| + \sqrt{\epsilon} \geq \sqrt{r^2 + \epsilon} \geq |\mu| - |y|$ for $r = y - \mu$.)

11
Convex Programming

11.1 Introduction

Our final chapter provides a concrete introduction to convex programming, interior point methods, and duality, three of the deepest and most pervasive themes of modern optimization theory. It takes considerable mathematical maturity to appreciate the subtlety of these topics, and it is impossible to do them justice in a short essay. Our philosophy here is to build on previous material on convexity and the MM algorithm. Exposing these connections reinforces old concepts while teaching new ones. The survey article [46] is especially recommended to readers wanting a more thorough introduction to the topics treated here.

Our first vignette describes adaptive barrier methods for convex programming. Barrier methods operate in the interior of the feasible region. An adaptive barrier method permits convergence to a boundary by gradually diminishing the strength of the corresponding barrier. The MM algorithm perspective suggests a way of accomplishing this while steadily decreasing the objective function. As our examples show, adaptive barrier methods have novel applications to linear and geometric programming. Our proof of convergence highlights the nice theoretical properties of the methods.

Specifying an interior feasible point is the first issue that must be faced in using a barrier method. Dykstra's algorithm finds the closest point to the intersection $\cap_{i=0}^{r-1} C_i$ of a finite number of closed convex sets. If C_i is defined by the convex constraint $h_i(x) \leq 0$, then one obvious tactic is to replace C_i by the set $C_i(\epsilon)$ defined by the constraint $h_j(x) \leq -\epsilon$ for some small number $\epsilon > 0$. Projecting onto the intersection of the $C_i(\epsilon)$ then produces an interior point.

Every convex program possesses a corresponding dual program, which can be simpler to solve than the original or primal program. We show how to construct dual programs and relate the absence of a duality gap to Slater's condition. We also point out important connections between duality and the Fenchel conjugate, bringing us full circle to the concrete introduction of Fenchel conjugates in Example 1.2.6 of Chapter 1.

The chapter closes with a brief introduction to linear classification and support vector machines. This material illustrates key ideas from duality and quadratic programming and gives us a chance to discuss yet another MM algorithm. Although we make no claims about our classification algorithm's computational efficiency, it does apply Dykstra's algorithm in an interesting way.

11.2 Adaptive Barrier Methods

The standard convex programming problem is to minimize a convex function $f(x)$ subject to affine equality constraints $a_{(i)}^* x - b_i = 0$ for $1 \leq i \leq p$ and concave inequality constraints $v_j(x) \geq 0$ for $1 \leq j \leq q$. We depart here from our previous convention on convex inequality constraints to avoid distracting negative signs in this section. In the logarithmic barrier method, we select some positive constant γ_m and minimize the modified objective function

$$h_m(x) = f(x) - \gamma_m \sum_{j=1}^{q} \ln v_j(x) \tag{11.1}$$

subject solely to the equality constraints. The presence of the barrier term $\gamma_m \ln v_j(x)$ keeps the constraint $v_j(x)$ inactive throughout the search provided that it starts inactive. Although the minimum point y of the original objective function $f(x)$ and the minimum point $y_{(m)}$ of the modified objective function $h_m(x)$ are not the same, the hope is that $y_{(m)}$ will tend to y as the barrier constant γ_m tends to 0.

Even though this hope is realized under fairly weak hypotheses, the necessity of iterations within iterations makes the barrier method too slow for practical use. One way of improving its speed is to change the barrier constant as the iterations proceed [16, 80]. This sounds vague, but matters simplify enormously if we view the construction of an adaptive barrier method from the perspective of the MM algorithm. Consider the following inequalities

$$
\begin{aligned}
& -v_j(x_{(m)}) \ln v_j(x) + v_j(x_{(m)}) \ln v_j(x_{(m)}) + dv_j(x_{(m)})(x - x_{(m)}) \\
\geq & -\frac{v_j(x_{(m)})}{v_j(x_{(m)})}[v_j(x) - v_j(x_{(m)})] + dv_j(x_{(m)})(x - x_{(m)}) \qquad (11.2) \\
= & -v_j(x) + v_j(x_{(m)}) + dv_j(x_{(m)})(x - x_{(m)}) \\
\geq & \ 0
\end{aligned}
$$

based on the concavity of the functions $\ln y$ and $v_j(x)$. Because equality holds throughout when $x = x_{(m)}$, we have identified a novel function majorizing 0 and incorporating a barrier for $v_j(x)$. (Such functions are known as Bregman distances in the literature [12].) The import of this discovery is that the surrogate function

$$
\begin{aligned}
g(x \mid x_{(m)}) &= f(x) - \gamma \sum_{j=1}^{q} v_j(x_{(m)}) \ln v_j(x) \qquad (11.3) \\
& \quad + \gamma \sum_{j=1}^{q} dv_j(x_{(m)})(x - x_{(m)})
\end{aligned}
$$

majorizes $f(x)$ up to an irrelevant additive constant. Here γ is a fixed positive constant. Minimization of the surrogate function drives $f(x)$ downhill while keeping the inequality constraints inactive. In the limit, one or more of the inequality constraints may become active.

Because minimization of the surrogate function $g(x \mid x_{(m)})$ cannot be accomplished in closed form, we must revert to the MM gradient algorithm. In performing one step of Newton's method, we need the first and second differentials

$$
\begin{aligned}
dg(x_{(m)} \mid x_{(m)}) &= df(x_{(m)}) \\
d^2g(x_{(m)} \mid x_{(m)}) &= d^2f(x_{(m)}) - \gamma \sum_{j=1}^{q} d^2v_j(x_{(m)}) \\
& \quad + \gamma \sum_{j=1}^{q} \frac{1}{v_j(x_{(m)})} \nabla v_j(x_{(m)}) dv_j(x_{(m)}).
\end{aligned}
$$

In view of the convexity of $f(x)$ and the concavity of the $v_j(x)$, it is obvious that $d^2g(x_{(m)} \mid x_{(m)})$ is positive semidefinite. If either $f(x)$ is strictly convex or the sum $\sum_{j=1}^{q} v_j(x)$ is strictly concave, then $d^2g(x_{(m)} \mid x_{(m)})$ is positive definite. As a safeguard in applying Newton's method, it is always a good idea to contract any proposed step so that $f(x_{(m+1)}) < f(x_{(m)})$ and $v_j(x_{(m+1)}) > \epsilon v_j(x_{(m)})$ for all j and a small ϵ such as 0.1.

The surrogate function (11.3) does not exhaust the possibilities for majorizing the objective function. If we replace the concave function $\ln y$ by the concave function $-y^{-\alpha}$ in our derivation (11.2), then we can construct for each $\alpha > 0$ and $\beta > 0$ the alternative surrogate

$$\begin{aligned} g(x \mid x_{(m)}) &= f(x) + \gamma \sum_{j=1}^{q} v_j(x_{(m)})^{\alpha+\beta} v_j(x)^{-\alpha} \\ &\quad + \gamma\alpha \sum_{j=1}^{q} v_j(x_{(m)})^{\beta-1} dv_j(x_{(m)})(x - x_{(m)}) \end{aligned} \tag{11.4}$$

majorizing $f(x)$ up to an irrelevant additive constant. This surrogate also exhibits an adaptive barrier that prevents the constraint $v_j(x)$ from becoming prematurely active. In this case, straightforward differentiation yields

$$\begin{aligned} dg(x_{(m)} \mid x_{(m)}) &= df(x_{(m)}) \\ d^2 g(x_{(m)} \mid x_{(m)}) &= d^2 f(x_{(m)}) - \gamma\alpha \sum_{j=1}^{q} v_j(x_{(m)})^{\beta-1} d^2 v_j(x_{(m)}) \\ &\quad + \gamma\alpha(\alpha+1) \sum_{j=1}^{q} v_j(x_{(m)})^{\beta-2} \nabla v_j(x_{(m)}) dv_j(x_{(m)}). \end{aligned}$$

Example 11.2.1 *A Geometric Programming Example*

In Example 5.4.6, we discussed how a geometric program can be transformed into a convex program. As a typical example, consider the problem of minimizing

$$f(x) = \frac{1}{x_1 x_2 x_3} + x_2 x_3$$

subject to

$$v(x) = 4 - 2x_1 x_3 - x_1 x_2 \geq 0$$

and positive values for the x_i. Making the change of variables $x_i = e^{y_i}$ transforms the problem into a convex program. With the choice $\gamma = 1$, the MM gradient algorithm with the exponential parameterization and the log surrogate (11.3) produces the iterates displayed in the top half of Table 11.1. In this case Newton's method performs well, and none of the safeguards is needed. The MM gradient algorithm with the power surrogate (11.4) does somewhat better. The results shown in the bottom half of Table 11.1 reflect the choices $\gamma = 1$, $\alpha = 1/2$, and $\beta = 1$. ■

In the presence of linear constraints, both updates for the adaptive barrier method rely on the quadratic approximation of the surrogate function $g(x \mid x_{(m)})$ using the calculated first and second differentials. This quadratic approximation is then minimized subject to the equality constraints as prescribed in Example 4.2.5.

TABLE 11.1. Solution of a Geometric Programming Problem

Iterates for the Log Surrogate

Iteration m	$f(x_{(m)})$	x_{m1}	x_{m2}	x_{m3}
1	2.0000	1.0000	1.0000	1.0000
2	1.7299	1.4386	0.9131	0.6951
3	1.6455	1.6562	0.9149	0.6038
4	1.5993	1.7591	0.9380	0.5685
5	1.5700	1.8256	0.9554	0.5478
10	1.5147	1.9614	0.9903	0.5098
15	1.5034	1.9910	0.9977	0.5023
20	1.5008	1.9979	0.9995	0.5005
25	1.5002	1.9995	0.9999	0.5001
30	1.5000	1.9999	1.0000	0.5000
35	1.5000	2.0000	1.0000	0.5000

Iterates for the Power Surrogate

Iteration m	$f(x_{(m)})$	x_{m1}	x_{m2}	x_{m3}
1	2.0000	1.0000	1.0000	1.0000
2	1.6478	1.5732	1.0157	0.6065
3	1.5817	1.7916	0.9952	0.5340
4	1.5506	1.8713	1.0011	0.5164
5	1.5324	1.9163	1.0035	0.5090
10	1.5040	1.9894	1.0011	0.5008
15	1.5005	1.9986	1.0002	0.5001
20	1.5001	1.9998	1.0000	0.5000
25	1.5000	2.0000	1.0000	0.5000

Example 11.2.2 *Linear Programming*

Consider the standard linear programming problem of minimizing c^*x subject to $Ax = b$ and $x \geq \mathbf{0}$. At iteration $m+1$ of the adaptive barrier method with the power surrogate (11.4), we minimize the quadratic approximation

$$c^*(x - x_{(m)}) + \tfrac{1}{2}\gamma\alpha(\alpha+1)\sum_{j=1}^{n} x_{mj}^{\beta-2}(x_j - x_{mj})^2$$

subject to $A(x - x_{(m)}) = \mathbf{0}$. According to Example 4.2.5, this minimization problem has solution

$$x_{(m+1)} = x_{(m)} - [D_{(m)}^{-1} - D_{(m)}^{-1}A^*(AD_{(m)}^{-1}A^*)^{-1}AD_{(m)}^{-1}]c,$$

where $D_{(m)}$ is a diagonal matrix with jth diagonal entry $\gamma\alpha(\alpha+1)x_{mj}^{\beta-2}$. It is convenient here to take $\gamma\alpha(\alpha+1) = 1$ and to step halve along the search direction $x_{(m+1)} - x_{(m)}$ whenever necessary. The case $\beta = 0$ bears a strong resemblance to Karmarkar's celebrated method of linear programming. ■

We now show that the MM algorithms based on the surrogates (11.3) and (11.4) converge under fairly natural conditions. In the interests of generality, we will not require that the objective function $f(x)$ be convex. However, we will retain the linear equality constraints $Ax = b$ and the concave inequality constraints $v_j(x) \geq 0$. To carry out this agenda, we assume that (a) $f(x)$ is continuously differentiable and coercive, (b) the constraint functions $v_j(x)$ are continuously differentiable, and (c) the sum $\sum_{j=1}^{q} v_j(x)$ is strictly concave. For simplicity, the objective functions and the constraint functions are defined throughout R^n. Either algorithm starts with a feasible point with all inequality constraints inactive.

For any subset $S \subset \{1, \ldots, q\}$, let M_S be the active manifold defined by the equalities $Ax = b$ and $v_j(x) = 0$ for $j \in S$ and the inequalities $v_j(x) > 0$ for $j \notin S$. If M_S is empty, then we can safely ignore it in the sequel. Let $P_S(x)$ denote the projection matrix satisfying $dv_j(x)P_S(x) = \mathbf{0}^*$ for every $j \in S$ and defined by

$$P_S(x) = I - dV_S(x)^*[dV_S(x)dV_S(x)^*]^{-1}dV_S(x), \qquad (11.5)$$

where $dV_S(x)$ consists of the row vectors $dv_j(x)$ with $j \in S$ stacked one atop another. For the matrix inverse appearing in equation (11.5) to make sense, the matrix $dV_S(x)$ should have full row rank. The matrix $P_S(x)$ projects a row vector onto the subspace perpendicular to the differentials $dv_j(x)$ of the active constraints. For reasons that will become clear later, we insist that $AP_S(x)$ have full row rank for any nonempty manifold M_S. When S is the empty set, we interpret $P_S(x)$ as the identity matrix I.

We will call a point $x \in M_S$ a stationary point if it satisfies the multiplier rule

$$df(x) + \lambda^* A - \sum_{j \in S} \mu_j dv_j(x) = \mathbf{0}^* \qquad (11.6)$$

for some vector λ and collection of nonnegative coefficients μ_j. According to Proposition 5.4.3, a stationary point furnishes a global minimum of $f(x)$ when $f(x)$ is convex. We will assume that each manifold M_S possesses at most a finite number of stationary points. This is certainly the case when $f(x)$ is strictly convex, but it can also hold for linear or even non-convex objective functions [38].

Proposition 11.2.1 *Under the conditions just sketched, the adaptive barrier algorithm based on either the surrogate function (11.3) or the surrogate function (11.4) with $\beta = 1$ converges to a stationary point of $f(x)$. If $f(x)$ is convex, then the algorithms converge to the unique global minimum y of $f(x)$ subject to the constraints.*

Proof: For the sake of brevity, we consider only the surrogate function (11.3). The coerciveness assumption guarantees that $f(x)$ possesses a minimum and that all iterates of a descent algorithm remain within a compact set. Because $g(x \mid x_{(m)})$ majorizes $f(x)$, it is coercive and attains its minimum value as well. Unless $f(x)$ is convex, the minimum point $x_{(m+1)}$ of $g(x \mid x_{(m)})$ may fail to be unique. If $f(x)$ is convex, then the concavity of the constraints $v_j(x)$ show that the surrogate function $g(x \mid x_{(m)})$ is convex. To demonstrate that it is strictly convex, consider the inequality

$$\begin{aligned}\sum_{j=1}^q v_j(x)\ln v_j(z) &\leq \sum_{j=1}^q \Big[v_j(x)\ln v_j(x) + v_j(z) - v_j(x)\Big] \\ &\leq \sum_{j=1}^q v_j(x)\ln v_j(x) + \Big[\sum_{j=1}^q dv_j(x)\Big](z-x)\end{aligned}$$

based on the concavity of $\ln t$. In view of the assumption that the sum $\sum_{j=1}^q v_j(x)$ is strictly concave, this inequality is strict whenever $z \neq x$. Hence, the barrier contributions force $g(x \mid x_{(m)})$ to be strictly convex and to possess a unique minimum point $x_{(m+1)}$ when $f(x)$ is convex.

Given these preliminaries, our attack is based on taking limits in the stationarity equation

$$\begin{aligned}\mathbf{0}^* &= df(x_{(m+1)}) + \lambda_{(m)}^* A \\ &\quad - \gamma \sum_{j=1}^q \left[\frac{v_j(x_{(m)})}{v_j(x_{(m+1)})} dv_j(x_{(m+1)}) - dv_j(x_{(m)})\right]\end{aligned} \tag{11.7}$$

and recovering the Lagrange multiplier rule. If we are to be successful in this regard, then we must show that

$$\lim_{m\to\infty} \|x_{(m+1)} - x_{(m)}\| = 0. \tag{11.8}$$

If the contrary is true, then there exists a subsequence $x_{(m_k)}$ such that

$$\liminf_{k\to\infty} \|x_{(m_k+1)} - x_{(m_k)}\| > 0.$$

Invoking compactness and passing to a subsubsequence if necessary, we can further assume that $\lim_{k\to\infty} x_{(m_k)} = u$ and $\lim_{k\to\infty} x_{(m_k+1)} = w$ with $u \neq w$. Exploiting the inequality $g(x_{(m+1)} \mid x_{(m)}) \leq g(x_{(m)} \mid x_{(m)})$ and the concavity of $v_j(x)$ and $\ln t$, we deduce the further inequalities

$$\begin{aligned} 0 &\leq \gamma \sum_{j=1}^q \big[v_j(x_{(m_k)}) - v_j(x_{(m_k+1)}) + dv_j(x_{(m_k)})(x_{(m_k+1)} - x_{(m_k)})\big] \\ &\leq \gamma \sum_{j=1}^q \left[v_j(x_{(m_k)}) \ln \frac{v_j(x_{(m_k)})}{v_j(x_{(m_k+1)})} + dv_j(x_{(m_k)})(x_{(m_k+1)} - x_{(m_k)})\right] \\ &= g(x_{(m_k+1)} \mid x_{(m_k)}) - f(x_{(m_k+1)}) - g(x_{(m_k)} \mid x_{(m_k)}) + f(x_{(m_k)}) \\ &\leq f(x_{(m_k)}) - f(x_{(m_k+1)}). \end{aligned}$$

Given that $f(x_{(m)})$ is bounded and decreasing, in the limit the difference $f(x_{(m_k)}) - f(x_{(m_k+1)})$ tends to 0. It follows that

$$\gamma \sum_{j=1}^{q} [v_j(u) - v_j(w) + dv_j(u)(w-u)] \quad = \quad 0,$$

contradicting the strict concavity of the sum $\sum_{j=1}^{q} v_j(x)$ and the hypothesis $u \neq w$. Hence, the limit (11.8) holds.

Because the iterates $x_{(m)}$ all belong to the same compact set, the proposition can be proved by demonstrating that every convergent subsequence $x_{(m_k)}$ converges to the same stationary point y. Consider such a subsequence with limit z. Let us divide the constraint functions $v_j(x)$ into those that are active at z and those that are inactive at z. In the former case, we take $j \in S$, and in the latter case, we take $j \in S^c$. In a moment we will demonstrate that z is a stationary point of M_S. Because by hypothesis there are only a finite number of manifolds M_s and only a finite number of stationary points per manifold, Proposition 10.3.1 implies that all limit points coincide and that the full sequence $x_{(m)}$ tends to a limit y. To finish the proof, we must demonstrate that y satisfies the multiplier rule for a constrained minimum. This last step is accomplished by taking limits in equality (11.7), assuming that $\lambda_{(m)}$ and the ratios $v_j(x_{(m)})/v_j(x_{(m+1)})$ all have limits. To avoid breaking the flow of our argument, we defer proof of this assumption. If $v_j(y) > 0$, then it is obvious that $v_j(x_{(m)})/v_j(x_{(m+1)})$ tends to 1, corresponding to a multiplier $\mu_j = 0$ for the inactive constraint j. If $v_j(y) = 0$, we must show that the limit of $v_j(x_{(m)})/v_j(x_{(m+1)})$ exceeds 1. But this is the case because $v_j(x_{(m+1)}) \leq v_j(x_{(m)})$ must hold for infinitely many m in order for $v_j(x_{(m)})$ to tend to 0.

We now return to the question of whether the limit z of the convergent subsequence $x_{(m_k)}$ is a stationary point. To demonstrate that the subsequence $\lambda_{(m_k)}$ converges, we multiply equation (11.7) on the right by the matrix $P_S(x_{(m_k+1)})A^*$ and solve for $\lambda_{(m_k)}$. This is possible because the matrix

$$\begin{aligned} B(x_{(m_k)}) &= AP_S(x_{(m_k+1)})A^* \\ &= AP_S(x_{(m_k+1)})P_S(x_{(m_k+1)})^*A^* \end{aligned}$$

has full rank by assumption. Since $P_S(x)$ annihilates $dv_j(x)$ for $j \in S$, and since $x_{(m_k+1)}$ converges to z, a brief calculation shows that

$$\lambda^* \quad = \quad \lim_{k\to\infty} \lambda_{(m_k)} \quad = \quad -df(z)P_S(z)A^*B(z)^{-1}.$$

To prove that the ratio $v_j(x_{(m_k)})/v_j(x_{(m_k+1)})$ has a limit for $j \in S$, we multiply equation (11.7) on the right by the matrix-vector product

$$P_{S_{-j}}(x_{(m_k+1)})\nabla v_j(x_{(m_k+1)}),$$

where $S_{-j} = S \setminus \{j\}$. This action annihilates all $dv_i(x_{(m_k+1)})$ with $i \in S_{-j}$ and makes it possible to express

$$\lim_{k\to\infty} \frac{v_j(x_{(m_k)})}{v_j(x_{(m_k+1)})} = \frac{[df(z) + \lambda^* A + \gamma dv_j(z)] P_{S_{-j}}(z) \nabla v_j(z)}{\gamma dv_j(z) P_{S_{-j}}(z) \nabla v_j(z)}.$$

Note that the denominator $\gamma dv_j(z) P_{S_{-j}}(z) \nabla v_j(z) > 0$ because $\gamma > 0$ and $dV_S(z)$ has full row rank. Given these results, we can legitimately take limits in equation (11.7) along the given subsequence and recover the multiplier rule (11.6) at z.

Now that we have demonstrated that $x_{(m)}$ tends to a unique limit y, we can show that $\lambda_{(m)}$ and the ratios $v_j(x_{(m)})/v_j(x_{(m+1)})$ tend to well-defined limits by the logic employed with the subsequence $x_{(m_k)}$. As noted earlier, this permits us to take limits in equation (11.7) and recover the multiplier rule at y. If $f(x)$ is convex, then y furnishes the global minimum. If there is another global minimum w, then the entire line segment between w and y consists of minimum points. This contradicts the assumption that there are at most a finite number of stationary points throughout the feasible region. Hence, the minimum point y is unique. ■

We will not undertake a systematic analysis of the rate of convergence of the adaptive barrier algorithms. The next example illustrates that the local rate of convergence can be linear even when one of the constraints $v_i(x) \geq 0$ is active at the minimum. Further partial results appear in [80].

Example 11.2.3 *Convergence for the Multinomial Distribution*

As pointed out in Example 1.4.2, the loglikelihood for a multinomial distribution reduces to $\sum_{i=1}^q n_i \ln p_i$, where n_i is the observed number of counts in category i and p_i is the probability attached to category i. Maximizing the loglikelihood subject to the constraints $p_i \geq 0$ and $\sum_{i=1}^q p_i = 1$ gives the explicit maximum likelihood estimates $p_i = n_i/n$ for n trials. To compute the maximum likelihood estimates iteratively using the surrogate function (11.3), we find a stationary point of the Lagrangian

$$-\sum_{i=1}^q n_i \ln p_i - \gamma \sum_{i=1}^q p_{mi} \ln p_i + \gamma \sum_{i=1}^q (p_i - p_{mi}) + \lambda\Big(\sum_{i=1}^q p_i - 1\Big).$$

Setting the ith partial derivative of the Lagrangian equal to 0 gives

$$-\frac{n_i}{p_i} - \frac{\gamma p_{mi}}{p_i} + \gamma + \lambda = 0. \tag{11.9}$$

Multiplying equation (11.9) by p_i, summing on i, and solving for λ yields $\lambda = n$. Substituting this value back in equation (11.9) produces

$$p_{m+1,i} = \frac{n_i + \gamma p_{mi}}{n + \gamma}.$$

At first glance it is not obvious that p_{mi} tends to n_i/n, but the algebraic rearrangement

$$\begin{aligned} p_{m+1,i} - \frac{n_i}{n} &= \frac{n_i + \gamma p_{mi}}{n+\gamma} - \frac{n_i}{n} \\ &= \frac{\gamma}{n+\gamma}\Big(p_{mi} - \frac{n_i}{n}\Big) \end{aligned}$$

shows that p_{mi} approaches n_i/n at the linear rate $\gamma/(n+\gamma)$. This is true regardless of whether $n_i/n = 0$ or $n_i/n > 0$. ■

11.3 Dykstra's Algorithm

In previous chapters, we have shown that the distance $d(x, C)$ from a point x to a set C is a uniformly continuous function of x. If the set C is closed and convex, then $d(x, C)$ is convex in x and $d(x, C) = \|P_C(x) - x\|$ for exactly one point $P_C(x) \in C$. Although it is generally impossible to calculate $P_C(x)$ explicitly, some specific projection operators are well known.

Example 11.3.1 *Examples of Projection Operators*

Closed Ball: If $C = \{y \in \mathsf{R}^n : \|y - z\| \le r\}$, then

$$P_C(x) = \begin{cases} z + r\frac{(x-z)}{\|x-z\|} & x \notin C \\ x & x \in C\,. \end{cases}$$

Closed Rectangle: If $C = [a, b]$ is a closed rectangle in R^n, then

$$P_C(x)_i = \begin{cases} a_i & x_i < a_i \\ x_i & x_i \in [a_i, b_i] \\ b_i & x_i > b_i\,. \end{cases}$$

Hyperplane: If $C = \{y \in \mathsf{R}^n : a^*y = b\}$ for $a \ne \mathbf{0}$, then

$$P_C(x) = x - \frac{a^*x - b}{\|a\|^2}a.$$

Closed Halfspace: If $C = \{y \in \mathsf{R}^n : a^*y \le b\}$ for $a \ne \mathbf{0}$, then

$$P_C(x) = \begin{cases} x - \frac{a^*x-b}{\|a\|^2}a & a^*x > b \\ x & a^*x \le b\,. \end{cases}$$

Subspace: If C is the range of a matrix A with full column rank, then

$$P_C(x) = A(A^*A)^{-1}A^*x.$$

Dykstra's algorithm [9, 30, 35] is designed to find the projection of a point x onto a finite intersection $\cap_{i=0}^{r-1} C_i$ of r closed convex sets. Here are some possible situations where Dykstra's algorithm applies.

Example 11.3.2 *Applications of Dykstra's Algorithm*

Linear Equalities: Any solution of the system of linear equations $Ax = b$ belongs to the intersection of the hyperplanes $a_{(i)}^* x = b_i$, where $a_{(i)}$ is the ith row of A.

Linear Inequalities: Any solution of the system of linear inequalities $Ax \leq b$ belongs to the intersection of the halfspaces $a_{(i)}^* x \leq b_i$, where $a_{(i)}$ is the ith row of A.

Isotone Regression: The least squares problem of minimizing the sum $\sum_{i=1}^n (x_i - w_i)^2$ subject to the constraints $w_i \leq w_{i+1}$ corresponds to projection of x onto the intersection of the halfspaces

$$C_i = \{w \in \mathsf{R}^n : w_i - w_{i+1} \leq 0\}, \quad 1 \leq i \leq n-1.$$

Convex Regression: The least squares problem of minimizing the sum $\sum_{i=1}^n (x_i - w_i)^2$ subject to the constraints $w_i \leq \frac{1}{2}(w_{i-1} + w_{i+1})$ corresponds to projection of x onto the intersection of the halfspaces

$$C_i = \Big\{w \in \mathsf{R}^n : w_i - \frac{1}{2}(w_{i-1} + w_{i+1}) \leq 0\Big\}, \quad 2 \leq i \leq n-1.$$

Quadratic Programming: To minimize the strictly convex quadratic form $\frac{1}{2}x^* Ax + b^* x + c$ subject to $Dx = e$ and $Fx \leq g$, we make the change of variables $y = A^{1/2}x$. This transforms the problem to one of minimizing

$$\begin{aligned} \frac{1}{2}x^* Ax + b^* x + c &= \frac{1}{2}\|y\|^2 + b^* A^{-1/2} y + c \\ &= \frac{1}{2}\|y + A^{-1/2} b\|^2 - \frac{1}{2} b^* A^{-1} b + c \end{aligned}$$

subject to $DA^{-1/2}y = e$ and $FA^{-1/2}y \leq g$. The solution in the y coordinates is determined by projecting $-A^{-1/2}b$ onto the convex feasible region determined by $DA^{-1/2}y = e$ and $FA^{-1/2}y \leq g$.

To state Dykstra's algorithm, it is helpful to label the closed convex sets $C_0, \ldots, C_{r-1}$ and denote their intersection by $C = \cap_{i=0}^{r-1} C_i$. The algorithm keeps track of a primary sequence $x_{(n)}$ and a companion sequence $e_{(n)}$. In the limit, $x_{(n)}$ tends to $P_C(x)$. To initiate the process, we set $x_{(-1)} = x$ and $e_{(-r)} = \cdots = e_{(-1)} = \mathbf{0}$. For $n \geq 0$ we then iterate via

$$\begin{aligned} x_{(n)} &= P_{C_{n \bmod r}}(x_{(n-1)} + e_{(n-r)}) \\ e_{(n)} &= x_{(n-1)} + e_{(n-r)} - x_{(n)}. \end{aligned}$$

Here $n \bmod r$ is the nonnegative remainder after dividing n by r. In essence, the algorithm cycles among the convex sets and projects the sum of the current vector and the relevant previous companion vector onto the current convex set. The proof that Dykstra's algorithm converges to $P_C(x)$ is not beyond us conceptually, but we omit it in the interests of brevity.

As an example, suppose $r = 2$, C_0 is the closed unit ball in R^2, and C_1 is the closed halfspace with $x_1 \geq 0$. The intersection C is the right half ball centered at the origin. Table 11.2 records the iterates of Dykstra's algorithm starting from the point $x = (-1, 2)$ and their eventual convergence to the geometrically obvious solution $(0, 1)$.

TABLE 11.2. Iterates of Dykstra's Algorithm

Iteration m	$\mathbf{x_{m1}}$	$\mathbf{x_{m2}}$
-1	-1.00000	2.00000
0	-0.44721	0.89443
1	0.00000	0.89443
2	-0.26640	0.96386
3	0.00000	0.96386
4	-0.14175	0.98990
5	0.00000	0.98990
10	-0.01814	0.99984
15	0.00000	0.99999
20	-0.00057	1.00000
25	0.00000	1.00000
30	-0.00002	1.00000
35	0.00000	1.00000

When C_i is a subspace, Dykstra's algorithm can dispense with the corresponding companion subsequence of $e_{(n)}$. In this case, $e_{(n)}$ is perpendicular to C_i whenever $n \bmod r = i$. Indeed, since $P_{C_i}(y)$ is a projection matrix, we have

$$\begin{aligned} x_{(n)} &= P_{C_i}(x_{(n-1)} + e_{(n-r)}) \\ &= P_{C_i}x_{(n-1)} + P_{C_i}e_{(n-r)} \\ &= P_{C_i}x_{(n-1)} \end{aligned}$$

under the perpendicularity assumption. The initial condition $e_{(i-r)} = \mathbf{0}$, the identity

$$\begin{aligned} e_{(n)} &= x_{(n-1)} - x_{(n)} + e_{(n-r)} \\ &= [I - P_{C_i}]x_{(n-1)} + e_{(n-r)} \\ &= P_{C_i^\perp} x_{(n-1)} + e_{(n-r)}, \end{aligned}$$

and induction show that $e_{(n)}$ belongs to the perpendicular complement $C_i^\perp$ if $n \bmod r = i$. When all of the C_i are subspaces, Dykstra's algorithm reduces to the method of alternating projections first studied by Von Neumann.

11.4 Dual Programs

The Lagrange multiplier rule summarizes much of what we know about minimizing $f(x)$ subject to the constraints $g_i(x) = 0$ for $1 \le i \le p$ and $h_j(x) \le 0$ for $1 \le j \le q$. Consequently, it is worth considering the standard Lagrangian function

$$L(x, \lambda, \mu) = f(x) + \sum_{i=1}^{p} \lambda_i g_i(x) + \sum_{j=1}^{q} \mu_j h_j(x)$$

with $\lambda_0 = 1$ in more detail. Here the multiplier vectors λ and μ are taken as arguments in addition to the variable x. For a convex program satisfying a constraint qualification such as Slater's condition, a global minimum $\hat{x}$ of $f(x)$ is also a global minimum of $L(x, \hat{\lambda}, \hat{\mu})$, where $\hat{\lambda}$ and $\hat{\mu}$ are the corresponding Lagrange multipliers. This fact follows from Proposition 5.4.3 because $\hat{x}$ is a stationary point of the convex function $L(x, \hat{\lambda}, \hat{\mu})$.

The behavior of $L(\hat{x}, \lambda, \mu)$ as a function of λ and μ is also interesting. Because $g_i(\hat{x}) = 0$ for all i and $\hat{\mu}_j h_j(\hat{x}) = 0$ for all j, we have

$$\begin{aligned} L(\hat{x}, \hat{\lambda}, \hat{\mu}) - L(\hat{x}, \lambda, \mu) &= \sum_{j=1}^{q} (\hat{\mu}_j - \mu_j) h_j(\hat{x}) \\ &= -\sum_{j=1}^{q} \mu_j h_j(\hat{x}) \\ &\ge 0. \end{aligned}$$

This proves the left inequality of the two saddle point inequalities

$$L(\hat{x}, \lambda, \mu) \le L(\hat{x}, \hat{\lambda}, \hat{\mu}) \le L(x, \hat{\lambda}, \hat{\mu}).$$

The left saddle point inequality immediately implies

$$\sup_{\lambda, \mu \ge \mathbf{0}} \inf_x L(x, \lambda, \mu) \le L(\hat{x}, \hat{\lambda}, \hat{\mu}) = f(\hat{x}).$$

By contrast, the right saddle point inequality entails

$$L(\hat{x},\hat{\lambda},\hat{\mu}) \;\le\; \inf_x L(x,\hat{\lambda},\hat{\mu}) \;\le\; \sup_{\lambda,\mu\ge \mathbf{0}} \inf_x L(x,\lambda,\mu).$$

Hence, we can recover the minimum value of $f(x)$ as

$$f(\hat{x}) \;=\; \sup_{\lambda,\mu\ge \mathbf{0}} \inf_x L(x,\lambda,\mu). \tag{11.10}$$

The problem of maximizing $\inf_x L(x,\lambda,\mu)$ is referred to as the dual program. It trades a potentially more complex objective function in the original, or primal, program for simpler constraints in the dual program. In a convex primal program, the equality constraint functions $g_i(x)$ are affine. If the inequality constraint functions $h_j(x)$ are also affine, then we can relate the dual program to the Fenchel conjugate

$$f^\star(y) \;=\; \sup_x\,[y^*x - f(x)] \tag{11.11}$$

of $f(x)$ introduced for functions of a real argument in Example 1.2.6. Suppose we write the constraints as $Vx = d$ and $Wx \le e$. Then the dual function equals

$$\begin{aligned}
\inf_x L(x,\lambda,\mu) &= \inf_x\,[f(x) + \lambda^*(Vx-d) + \mu^*(Wx-e)] \\
&= -\lambda^*d - \mu^*e + \inf_x[f(x) + (V^*\lambda + W^*\mu)^*x] \\
&= -\lambda^*d - \mu^*e - f^\star(-V^*\lambda - W^*\mu).
\end{aligned}$$

It may be that $f^\star(y)$ equals ∞ for certain values of y, but we can ignore these values in maximizing $\inf_x L(x,\lambda,\mu)$. The calculation

$$\begin{aligned}
&\sup_x\,\{[\alpha y + (1-\alpha)z]^*x - f(x)]\} \\
=\;& \sup_x\,[\alpha y^*x - \alpha f(x) + (1-\alpha)z^*x - (1-\alpha)f(x)] \\
\le\;& \alpha \sup_x\,[y^*x - f(x)] + (1-\alpha)\sup_x\,[z^*x - f(x)]
\end{aligned}$$

shows that the Fenchel conjugate is a convex function.

Equality (11.10) is predicated on Slater's condition or some substitute that forces the preferred form of the multiplier rule at a minimum of $f(x)$ subject to the constraints. In the absence such a guarantee, we can still recover a weak form of duality based on the identity

$$f(\hat{x}) \;=\; \inf_x \sup_{\lambda,\mu\ge \mathbf{0}} L(x,\lambda,\mu),$$

which stems from the fact

$$\sup_{\lambda,\mu\ge \mathbf{0}} L(x,\lambda,\mu) \;=\; \begin{cases} f(x) & x \text{ feasible} \\ \infty & x \text{ infeasible.} \end{cases}$$

Because $\inf_x L(x, \lambda, \mu) \leq L(\hat{x}, \lambda, \mu) \leq f(\hat{x})$, we can assert that

$$\sup_{\lambda,\mu \geq \mathbf{0}} \inf_x L(x, \lambda, \mu) \;\leq\; f(\hat{x}) \;=\; \inf_x \sup_{\lambda,\mu \geq \mathbf{0}} L(x, \lambda, \mu).$$

In other words, the minimum value of the primal problem exceeds the maximum value of the dual problem. Slater's condition guarantees that there is no duality gap when the primal problem has a finite minimum. Weak duality also makes it evident that if the primal program has no minimum, then the dual program has no feasible point, and that if the dual program has no maximum, then the primal program has no feasible point.

Example 11.4.1 *The Dual of a Linear Program*

The standard linear program minimizes $f(x) = z^*x$ subject to the linear equality constraints $Vx = d$ and the nonnegativity constraints $x \geq \mathbf{0}$. It is obvious that $f^\star(y) = \infty$ unless $y = z$, in which case $f^\star(z) = 0$. Thus, the dual program maximizes $-\lambda^* d$ subject to the constraints $-V^*\lambda + \mu = z$ and $\mu \geq \mathbf{0}$. In this situation, arguments based on the simplex method of linear programming show that there is no duality gap when either the primal or dual programs has a finite solution [38, 89]. ■

Example 11.4.2 *The Dual of a Strictly Convex Quadratic Program*

We have repeatedly visited the problem of minimizing the strictly convex function $f(x) = \frac{1}{2}x^*Ax + b^*x + c$ subject to the linear equality constraints $Vx = d$ and the linear inequality constraints $Wx \leq e$. Ignoring the constraints, $f(x)$ achieves its minimum value $-\frac{1}{2}b^*A^{-1}b + c$ at the point $x = -A^{-1}b$. Analogous considerations imply that $f(x)$ has Fenchel conjugate

$$f^\star(y) \;=\; \frac{1}{2}(y - b)^*A^{-1}(y - b) - c.$$

If there are no equality constraints, then the dual program maximizes the quadratic

$$\begin{aligned}
\inf_x L(x, \mu) &= -\mu^*e - f^\star(-W^*\mu) \\
&= -\mu^*e - \frac{1}{2}(W^*\mu + b)^*A^{-1}(W^*\mu + b) + c \\
&= -\frac{1}{2}\mu^*WA^{-1}W^*\mu - (e + WA^{-1}b)^*\mu - \frac{1}{2}b^*A^{-1}b + c
\end{aligned}$$

subject to $\mu \geq \mathbf{0}$. The dual program is substantially simpler than the primal program, particularly when the number of inequality constraints is small. In the presence of equality constraints, the maximum of the function

$\inf_x L(x, \lambda, \mu) = -\lambda^* d - \mu^* e - f^\star(-V^*\lambda - W^*\mu)$ with respect to λ occurs where

$$\begin{aligned} \mathbf{0} &= -d + V\nabla f^\star(-V^*\lambda - W^*\mu) \\ &= -d + VA^{-1}(-V^*\lambda - W^*\mu - b). \end{aligned}$$

If V has full row rank, then this equation has solution

$$\lambda = -(VA^{-1}V^*)^{-1}(VA^{-1}W^*\mu + VA^{-1}b + d).$$

The corresponding maximum is a messy quadratic in μ. The dual program trades a more complicated quadratic for elimination of the equality constraints and simplification of the inequality constraints to $\mu \geq \mathbf{0}$. ■

Example 11.4.3 *Duffin's Counterexample*

Consider the convex program of minimizing $f(x) = e^{-x_2}$ subject to the inequality constraint $h(x) = \|x\| - x_1 \leq 0$ on R^2. This problem does not satisfy Slater's condition because all feasible x satisfy $x_2 = 0$ and consequently $h(x) = 0$ and $f(x) = 1$. To demonstrate that there is a duality gap, we show that the dual function

$$\inf_x L(x, \mu) = \inf_x [e^{-x_2} + \mu(\|x\| - x_1)]$$

is identically 0. Because $L(x, \mu) \geq 0$ for all x and $\mu \geq 0$, it suffices to prove that $L(x, \mu)$ can be made less than any positive ϵ. Choose an x_2 so that $e^{-x_2} < \epsilon/2$. Having chosen x_2, we choose x_1 so that

$$\begin{aligned} \sqrt{x_1^2 + x_2^2} - x_1 &= x_1\sqrt{1 + \frac{x_2^2}{x_1^2}} - x_1 \\ &\leq x_1\left(1 + \frac{x_2^2}{2x_1^2}\right) - x_1 \\ &= \frac{x_2^2}{2x_1} \\ &< \frac{\epsilon}{2\mu}. \end{aligned}$$

Thus, the minimum value 1 of the primal problem is strictly greater than the maximum value 0 of the dual problem. ■

Example 11.4.4 *Fenchel Biconjugate*

The Fenchel conjugate provides one of the loveliest manifestations of duality [7]. The inequality $f^\star(y) \geq y^*x - f(x)$ follows directly from the definition of $f^\star(y)$ and implies the further inequality $f(x) \geq x^*y - f^\star(y)$. Taking the supremum over all y shows that $f(x) \geq f^{\star\star}(x)$ for all x. The reverse inequality is true at a point z whenever $f(x)$ possess a subdifferential $df(z)$ there. Indeed, the choice $y = df(z)$ shows that

$$\sup_y \inf_x [f(x) - f(z) - y^*(x - z)] \geq 0.$$

However,

$$\begin{aligned}
\sup_y \inf_x [f(x) - f(z) - y^*(x - z)] &= \sup_y \{y^*z - f(z) + \inf_x [f(x) - y^*x]\} \\
&= \sup_y \{y^*z - f(z) - \sup_x [y^*x - f(x)]\} \\
&= \sup_y [y^*z - f(z) - f^\star(y)] \\
&= f^{\star\star}(z) - f(z),
\end{aligned}$$

proving that $f^{\star\star}(z) \geq f(z)$ and hence that $f^{\star\star}(z) = f(z)$. ■

11.5 Linear Classification

Classification is a commonly encountered task in statistics. The binary classification problem can be phrased in terms of a training sequence of observation vectors $v_{(1)}, \ldots, v_{(m)}$ from R^n and an associated sequence of population indicators $s_1, \ldots, s_n$ from $\{-1, +1\}$. In favorable situations, the two different populations can be separated by a hyperplane defined by a unit vector z and constants $c_1 \leq c_2$ in the sense that

$$\begin{aligned}
z^* v_{(i)} &\leq c_1, \quad s_i = -1 \\
z^* v_{(i)} &\geq c_2, \quad s_i = +1.
\end{aligned}$$

The optimal separation occurs when the difference $c_2 - c_1$ is maximized. The linear classification problem can be simplified by rewriting the separation conditions as

$$\begin{aligned}
z^* v_{(i)} - \frac{c_1 + c_2}{2} &\leq -\frac{c_2 - c_1}{2}, \quad s_i = -1 \\
z^* v_{(i)} - \frac{c_1 + c_2}{2} &\geq +\frac{c_2 - c_1}{2}, \quad s_i = +1.
\end{aligned}$$

If we let $a = (c_2 - c_1)/2$, $b = (c_1 + c_2)/(c_2 - c_1)$, and $y = a^{-1}z$, then these become

$$\begin{aligned}
y^* v_{(i)} - b &\leq -1, \quad s_i = -1 \\
y^* v_{(i)} - b &\geq +1, \quad s_i = +1.
\end{aligned} \tag{11.12}$$

Thus, the linear classification problem reduces to one of minimizing $\frac{1}{2}\|y\|^2$ subject to the inequality constraints (11.12). Observe that the constraint functions are linear in the parameter vector $(y^*, b)^*$ in this semidefinite quadratic programming problem. Unfortunately, because the component b is not involved in $\frac{1}{2}\|y\|^2$, Dykstra's algorithm does not apply. Once we find y and b, we can classify a new test vector v in the $s = -1$ population when $y^*v - b < 0$ and in the $s = +1$ population when $y^*v - b > 0$.

There is no guarantee that a feasible vector $(y^*, b)^*$ exists for the linear classification problem as stated. A more realistic version of the problem imposes the inequality constraints

$$s_i(y^* v_{(i)} - b) \quad \geq \quad 1 - \epsilon_i \tag{11.13}$$

using a slack variable $\epsilon_i \geq 0$. To penalize deviation from the ideal of perfect separation by a hyperplane, we modify the objective function to be

$$f(y, b, \epsilon) \quad = \quad \frac{1}{2}\|y\|^2 + \delta \sum_{i=1}^{m} \epsilon_i \tag{11.14}$$

for some tuning constant $\delta > 0$. The constraints (11.13) and $\epsilon_i \geq 0$ are again linear in the parameter vector $x = (y^*, b, \epsilon^*)^*$. The Lagrangian

$$\begin{aligned} L(y, b, \epsilon, \mu) \quad &= \quad \frac{1}{2}\|y\|^2 + \delta \sum_{i=1}^{m} \epsilon_i - \sum_{i=1}^{m} \mu_{m+i}\epsilon_i \\ &\quad + \sum_{i=1}^{m} \mu_i[-s_i(v_{(i)}^* y - b) + 1 - \epsilon_i] \end{aligned}$$

involves $2m$ nonnegative multipliers $\mu_1, \ldots, \mu_{2m}$.

In most instances, it is simpler to solve the dual problem rather than the primal problem. We can formulate the dual by following the steps of Example 11.4.2. The transpose W^* of the inequality matrix W and the vector e of that example turn out to be

$$W^* \quad = \quad -\begin{pmatrix} s_1 v_{(1)} & \cdots & s_m v_{(m)} & \mathbf{0} \\ s_1 & \cdots & s_m & 0 \\ & I_m & & I_m \end{pmatrix}, \qquad e \quad = \quad \begin{pmatrix} -\mathbf{1} \\ \mathbf{0} \end{pmatrix}$$

in the current setting. It is easy to check that the Fenchel conjugate is

$$\begin{aligned} f^\star(p, q, r) \quad &= \quad \sup_{(y^*, b, \epsilon^*)^*} \Big[p^* y + qb + r^*\epsilon - \frac{1}{2}\|y\|^2 - \delta \sum_{i=1}^{m} \epsilon_i\Big] \\ &= \quad \begin{cases} \infty, & q \neq 0 \text{ or } r_i > \delta \text{ for some } i \\ p^*p - \frac{1}{2}\|p\|^2, & \text{otherwise} \end{cases} \\ &= \quad \begin{cases} \infty, & q \neq 0 \text{ or } r_i > \delta \text{ for some } i \\ \frac{1}{2}\|p\|^2, & \text{otherwise.} \end{cases} \end{aligned}$$

The dual problem of maximizing $-\mu^*e - f^\star(-W^*\mu)$ subject to the constraints $\mu_i \geq 0$ is further constrained by the infinite values of $f^\star(-W^*\mu)$.

The finiteness condition $q = 0$ in the definition of the Fenchel conjugate implies the constraint $\sum_{i=1}^m s_i\mu_i = 0$. The finiteness condition $r_i \le \delta$ implies the constraint $\mu_i \le \delta$. For the vector p we substitute the linear combination $\sum_{i=1}^m s_i\mu_i v_{(i)}$. Hence, the dual problem consists in maximizing

$$\begin{aligned}
-\mu^* e - f^\star(-W^*\mu) &= \sum_{i=1}^m \mu_i - \frac{1}{2}\Big\|\sum_{i=1}^m s_i\mu_i v_{(i)}\Big\|^2 \\
&= \sum_{i=1}^m \mu_i - \frac{1}{2}\sum_{i=1}^m\sum_{j=1}^m s_i\mu_i v_{(i)}^* v_{(j)} s_j\mu_j
\end{aligned}$$

subject to the constraints $\sum_{i=1}^m s_i\mu_i = 0$ and $0 \le \mu_i \le \delta$ for all $1 \le i \le m$.

The solution of the dual problem immediately yields the solution of the primal problem. For instance, the Lagrangian conditions

$$\frac{\partial}{\partial y_j} L(y, b, \epsilon, \mu) = 0$$

give $y = \sum_{i=1}^m \mu_i s_i v_{(i)}$. The Lagrangian condition

$$\begin{aligned}
\frac{\partial}{\partial \epsilon_j} L(y, b, \epsilon, \mu) &= \delta - \mu_j - \mu_{m+j} \\
&= 0
\end{aligned}$$

implies that $\mu_{m+j} = \delta - \mu_j$. The multiplier conditions

$$\begin{aligned}
0 &= \mu_i[-s_i(v_{(i)}^* y - b) + 1 - \epsilon_i] \\
0 &= \mu_{m+j}\epsilon_j
\end{aligned}$$

can be used to determine b and ϵ_i for $1 \le i \le m$. If we choose an index j such that $0 < \mu_j < \delta$, then $\mu_{m+j} > 0$ and $\epsilon_j = 0$. It follows that $-s_j(v_{(j)}^* y - b) + 1 = 0$ and that $b = v_{(j)}^* y - s_j$ since $s_j = \pm 1$. Given b, all ϵ_j with $\mu_j > 0$ are determined. If $\mu_j = 0$, then $\mu_{m+j} > 0$ and $\epsilon_j = 0$.

Despite these interesting maneuvers, we have not actually shown how to solve the dual problem. For linear classification problems with many training vectors $v_{(1)}, \ldots, v_{(m)}$ in a high-dimensional space R^n, it is imperative to formulate an efficient algorithm. We now derive an MM algorithm that invokes Dykstra's algorithm at every iteration. First, we restate the dual problem as finding the minimum of the quadratic

$$g(\mu) = \frac{1}{2}\sum_{i=1}^m\sum_{j=1}^m s_i\mu_i v_{(i)}^* v_{(j)} s_j\mu_j - \sum_{i=1}^m \mu_i$$

subject to the constraints. If the vectors $v_{(i)}$ are linearly independent, then $g(\mu)$ is strictly convex. This fact follows immediately from the identity

$$\sum_{i=1}^m\sum_{j=1}^m s_i\mu_i v_{(i)}^* v_{(j)} s_j\mu_j = \Big\|\sum_{i=1}^m s_i\mu_i v_{(i)}\Big\|^2.$$

Another way of summarizing the situation is to say that the symmetric matrix M with entries $m_{ij} = s_i v_{(i)}^* v_{(j)} s_j$ is positive definite.

To majorize $g(\mu)$, we note that the largest eigenvalue λ of M satisfies $\mu^* M \mu \leq \lambda \|\mu\|^2$ for all μ. If we complete the square and use this bound, then we find that

$$\begin{aligned} g(\mu) &= \frac{1}{2}\mu^* M \mu - \mathbf{1}^* \mu \\ &= \frac{1}{2}(\mu - \mu_{(k)})^* M (\mu - \mu_{(k)}) + (M\mu_{(k)} - \mathbf{1})^* \mu - \frac{1}{2}\mu_{(k)}^* M \mu_{(k)} \\ &\leq \frac{1}{2}\lambda(\mu - \mu_{(k)})^*(\mu - \mu_{(k)}) + (M\mu_{(k)} - \mathbf{1})^* \mu - \frac{1}{2}\mu_{(k)}^* M \mu_{(k)} \\ &= \frac{1}{2}\lambda \mu^* \mu + (M\mu_{(k)} - \mathbf{1} - \lambda \mu_{(k)})^* \mu + \frac{1}{2}\lambda \mu_{(k)}^* \mu_{(k)} - \frac{1}{2}\mu_{(k)}^* M \mu_{(k)}. \end{aligned}$$

We now complete the square again and construct the majorizing function

$$g(\mu) \leq \frac{1}{2}\lambda \|\mu - w_{(k)}\|^2 + c_k = h(\mu \mid \mu_{(k)}),$$

where $w_{(k)}$ is the vector $\lambda^{-1}(\mathbf{1} + \lambda\mu_{(k)} - M\mu_{(k)})$ and c_k is a scalar that does not depend on μ. We can minimize $h(\mu \mid \mu_{(k)})$ by finding the closest point μ to $w_{(k)}$ in the convex set defined by the constraints $\sum_{i=1}^m s_i \mu_i = 0$ and $0 \leq \mu_i \leq \delta$ for all $1 \leq i \leq m$. Because the constraints are so simple, Dykstra's algorithm is ideal for minimizing $h(\mu \mid \mu_{(k)})$. In fact, there is no need to run Dykstra's algorithm to completion because all we must do is make progress in reducing $h(\mu \mid \mu_{(k)})$.

This hybrid MM-Dykstra algorithm is predicated on knowing the largest eigenvalue λ of M. A fast algorithm for finding λ that uses only matrix times vector multiplication is sketched in Problem 10 of Chapter 9. Finding all of the eigenvalues of M is naturally much harder. The reference [110] discusses other techniques for solving the dual problem. Some of the special tactics mentioned there apply to the MM-Dykstra algorithm as well. For instance, the hybrid algorithm converges very slowly for large m because the majorization $\mu^* M \mu \leq \lambda \|\mu\|^2$ is too crude. Thus, it can be advantageous to apply the hybrid algorithm to varying subsets of the μ_i while holding the remaining μ_i fixed.

11.6 Support Vector Machines

The rather odd phrase "support vector machines" refers to a body of statistical and computational techniques that build on the material developed in the last section [56, 110, 119]. The modern theory of machine learning makes a distinction between pattern space, where observations are actually taken, and feature space, where two or more populations can be linearly separated based on their mapped features. Pattern space may include purely qualitative distinctions such as gender as well as quantitative measurements such as cholesterol and blood pressure levels. Feature space has an explicit geometry derived from an inner product. To classify a pattern y, we map it into feature space via some transformation $\Phi(y)$ that combines the different components of y in many informative ways. Feature space is typically high-dimensional or even-infinite dimensional. The inner product $k(x,y) = \langle \Phi(x), \Phi(y) \rangle$ of two mapped patterns x and y is called a kernel. Any finite set of patterns $y_{(1)}, \ldots, y_{(m)}$ generates a positive definite matrix $M = (k(y_{(i)}, y_{(j)})$ provided the corresponding features $v_{(1)} = \Phi(y_{(1)})$ through $v_{(m)} = \Phi(y_{(n)})$ are linearly independent vectors. To prove this fact, we simply note that

$$w^* M w = \Big\| \sum_{i=1}^{m} w_i v_{(i)} \Big\|^2$$

in the norm associated with the inner product in feature space.

Example 11.6.1 *Kronecker Products*

Suppose the vectors u and v have m and n entries, respectively. Their Kronecker product is the vector

$$u \otimes v = \begin{pmatrix} u_1 v \\ u_2 v \\ \vdots \\ u_m v \end{pmatrix} = \begin{pmatrix} u_1 v_1 \\ \vdots \\ u_m v_n \end{pmatrix}$$

whose entries correspond to all possible products $u_i v_j$. From this definition, one can easily demonstrate the inner product rule

$$(u \otimes v)^*(w \otimes x) = (u^* w)(v^* x).$$

The map

$$\Phi(y) = \begin{pmatrix} y \\ 1 \end{pmatrix} \otimes \begin{pmatrix} y \\ 1 \end{pmatrix}$$

encodes all possible entries y_i and their products $y_i y_j$ and gives rise to the kernel $k(x,y) = \Phi(x)^*\Phi(y) = (x^*y+1)^2$. This example generalizes to r-fold Kronecker products and generates the kernel $k(x,y) = (x^*y+1)^r$ encoding all possible monomials $y_{i_1} \cdots y_{i_s}$ with $s \le r$. ■

Example 11.6.2 *Characteristic Functions*

Let Z be a random vector with n components and probability density $f(x)$. A complex-valued function $g(Z)$ of Z is said to be square-integrable provided $\mathrm{E}[|g(Z)|^2] < \infty$. The set $\mathrm{L}^2(Z)$ of square-integrable functions is an infinite-dimensional inner product space with inner product

$$\langle g, h \rangle \quad = \quad \mathrm{E}[g(Z)\bar{h}(Z)],$$

where $\bar{h}(Z)$ denotes the complex conjugate of $h(Z)$. Dealing with complex-valued functions allows us to define the map $\Phi(x) = e^{ix^*Z}$, $i = \sqrt{-1}$, from R^n into $\mathrm{L}^2(Z)$. This map generates the translation-invariant kernel

$$\begin{aligned} \langle \Phi(x), \Phi(y) \rangle &= \mathrm{E}\left[e^{i(x-y)^*Z}\right] \\ &= \hat{f}(x-y), \end{aligned}$$

which is the characteristic function $\hat{f}(z)$ of Z. There are many such characteristic functions. A prime example is the Gaussian function $\hat{f}(z) = e^{-c\|z\|^2}$ for $c > 0$ discussed in the Appendix. If the underlying random vector Z and its negative $-Z$ are identically distributed, then the characteristic function $\hat{f}(z)$ is real valued. ■

The primary virtue of combining pattern and feature spaces is that it enables nonlinear classification. The map $\Phi(y)$ transforming a pattern into its features is almost always nonlinear. However, once we have constructed the features $v_{(i)} = \Phi(y_{(i)})$ corresponding to a set of training patterns $y_{(i)}$, we can find an optimal hyperplane separating them into two populations by purely linear means. Everything said in the previous section continues to hold provided we substitute the kernel values $k(y_{(i)}, y_{(j)})$ for the corresponding inner products $v_{(i)}^* v_{(j)}$. Indeed, it is safe to say in most applications that the map $\Phi(y)$ and the full complexity of feature space fade into the shadows. The kernel $k(x, y)$ they leave behind suffices for practical purposes.

11.7 Problems

1. Show that the surrogate function (11.4) majorizes $f(x)$.

2. Prove rigorously that the adaptive barrier algorithms have found the global minimum of the geometric programming problem in Table 11.1.

3. If C is a closed convex set in R^n, then show that

$$[y - P_C(x)]^*[x - P_C(x)] \quad \leq \quad 0 \tag{11.15}$$

for every $y \in C$. Furthermore, argue that $P_C[P_C(x)] = P_C(x)$ and

$$\|P_C(x) - P_C(y)\| \quad \leq \quad \|x - y\|.$$

(Hints: See Proposition 5.2.2. Also write

$$\begin{aligned}\|x-y\|^2 &= \|P_C(x)-P_C(y)\|^2+\|x-P_C(x)-y+P_C(y)\|^2\\ &\quad -2[P_C(y)-P_C(x)]^*[x-P_C(x)]\\ &\quad -2[P_C(x)-P_C(y)]^*[y-P_C(y)]\end{aligned}$$

and use inequality (11.15).)

4. If C is a closed convex set in R^n and $x \notin C$, then demonstrate that

$$d(x,C) = \inf_{y\in C}\sup_{\|z\|=1} z^*(x-y) = \sup_{\|z\|=1}\inf_{y\in C} z^*(x-y).$$

Also prove that there exists a unit vector z with

$$d(x,C) = \inf_{y\in C} z^*(x-y).$$

(Hints: The first equality follows from the Cauchy-Schwarz inequality and the definition of $d(x,C)$. The rest of the problem depends on the Cauchy-Schwarz inequality, the particular choice

$$z = d(x,C)^{-1}[x-P_C(x)],$$

and inequality (11.15).)

5. Let S and T be subspaces of R^n. Demonstrate that the projections P_S and P_T satisfy $P_SP_T = P_{S\cap T}$ if and only if $P_SP_T = P_TP_S$.

6. Suppose C is a closed convex set. If y is on the line segment between x and $P_C(x)$, then prove that $P_C(y) = P_C(x)$.

7. Let C be a closed convex set in R^n. Show that

   (a) $d(x+y, C+y) = d(x,C)$ for all x and y.

   (b) $P_{C+y}(x+y) = P_C(x)+y$ for all x and y.

   (c) $d(ax, aC) = |a|d(x,C)$ for all x and real a.

   (d) $P_{aC}(ax) = aP_C(x)$ for all x and real a.

   Let S be a subspace of R^n. Show that

   (a) $d(x+y,S) = d(x,S)$ for all $x \in \mathsf{R}^n$ and $y \in S$.

   (b) $P_S(x+y) = P_S(x)+y$ for all $x \in \mathsf{R}^n$ and $y \in S$.

   (c) $d(ax,S) = |a|d(x,S)$ for all $x \in \mathsf{R}^n$ and real a.

   (d) $P_S(ax) = aP_S(x)$ for all $x \in \mathsf{R}^n$ and real a.

TABLE 11.3. Properties of the Fenchel Conjugate

f	**f***
$f(Mx)$	$f^\star[(M^*)^{-1}y]$
$f(x-v)$	$f^\star(y)+v^*y$
$f(x)-v^*x$	$f^\star(y+v)$
$af(x)$	$af^\star(y/a)$
$\sum_{i=1}^n f_i(x_i)$	$\sum_{i=1}^n f_i^\star(y_i)$

8. Show that the Fenchel conjugate defined by equation (11.11) satisfies the properties displayed in Table 11.3, where M is an $n \times n$ matrix, v is an $n \times 1$ vector, and a is a positive scalar.

9. Suppose that the convex function $g(x)$ satisfies $g(x) \leq f(x)$ for all x. Prove that $g(x) \leq f^{\star\star}(x) \leq f(x)$ for all x. Use this fact to show that the function $f(x) = (x^2-1)^2$ has Fenchel biconjugate

$$f^{\star\star}(x) = \begin{cases} 0 & |x| \leq 1 \\ (x^2-1)^2 & |x| > 1 \end{cases}$$

and that the function

$$f(x) = \begin{cases} |x| & |x| \leq 1 \\ 2-|x| & 1 < |x| \leq 3/2 \\ |x|-1 & |x| > 3/2 \end{cases}$$

has Fenchel biconjugate

$$f^{\star\star}(x) = \begin{cases} \frac{|x|}{3} & |x| \leq 3/2 \\ f(x) & |x| > 3/2 \,. \end{cases}$$

10. Demonstrate that $f(x) = \frac{1}{2}\|x\|^2$ is the only function satisfying the identity $f(x) = f^\star(x)$ for all x. (Hint: Use the trivial inequality $f^\star(y) + f(x) \geq y^*x$ to prove that $f(x) \geq \frac{1}{2}\|x\|^2$ when $f(x) = f^\star(x)$. For the reverse inequality, substitute this inequality in the definition of $f^\star(y)$.)

11. Let $f(y)$ be a differentiable function from R^n to R. Prove that x is a global minimum of $f(y)$ if and only if $\nabla f(x) = \mathbf{0}$ and $f^{\star\star}(x) = f(x)$ [62]. (Hints: If x is a global minimum, then $\mathbf{0}^*$ is a subdifferential of $f(y)$ at x. Conversely, if the two conditions hold, then show that every directional derivative $d_v f(x)$ satisfies $d_v f^{\star\star}(x) \leq 0$. Because $-d_{-v} f^{\star\star}(x) \leq d_v f^{\star\star}(x)$, we have in fact $d_v f^{\star\star}(x) = 0$ for every direction v. Now use Problem 13 of Chapter 3 to establish that $\nabla f^{\star\star}(x) = \mathbf{0}$. Because $f^{\star\star}(y)$ is convex, x minimizes $f^{\star\star}(y)$.)

TABLE 11.4. Some Specific Fenchel Conjugates

$\mathbf{f(x)}$	$\mathbf{f^\star(y)}$
x	$\begin{cases} 0 & y = 1 \\ \infty & y \neq 1 \end{cases}$
$\|x\|$	$\begin{cases} 0 & \|y\| \leq 1 \\ \infty & \|y\| > 1 \end{cases}$
$\begin{cases} -\ln x & x > 0 \\ \infty & x \leq 0 \end{cases}$	$\begin{cases} -1 - \ln(-y) & y < 0 \\ \infty & y \geq 0 \end{cases}$
e^x	$\begin{cases} y \ln y - y & y > 0 \\ 0 & y = 0 \\ \infty & y < 0 \end{cases}$

12. Derive the Fenchel conjugates displayed in Table 11.4 for functions of a single real variable.

13. Let $f(x)$ be a strictly convex function on the real line. Prove that $f'(x)$ and $[f^\star(y)]'$ are functional inverses when $f(x)$ and its derivatives are sufficiently well behaved.

14. Consider the convex programming problem of minimizing the convex function $f(x)$ subject to the affine equality constraints $g_i(x) = 0$ for $1 \leq i \leq p$ and the convex inequality constraints $h_j(x) \leq c_j$ for $1 \leq j \leq q$. Show that the solution is a convex function of the vector $c = (c_1, \ldots, c_q)^*$.

15. Let A and B be positive semidefinite matrices of the same dimension. Show that the matrix $aA + bB$ is positive semidefinite for every pair of nonnegative scalars a and b. Thus, the set of kernel functions is a cone.

16. The Hadamard product $C = A \circ B$ of two matrices $A = (a_{ij})$ and $B = (b_{ij})$ of the same dimensions has entries $c_{ij} = a_{ij}b_{ij}$. If A and B are positive semidefinite matrices, then show that $A \circ B$ is positive semidefinite. This proves that the pointwise product of two kernel functions is a kernel function. (Hint: Let X and Y be independent multivariate normal random vectors with mean $\mathbf{0}$ and covariance matrices A and B. Show that the vector Z with entries $Z_i = X_iY_i$ has covariance matrix $A \circ B$.)

17. Suppose the power series $f(t) = \sum_{n=0}^{\infty} c_n t^n$ has nonnegative coefficients and converges for all t. Demonstrate that the $m \times m$ matrix with entry $f(v_{(i)}^* v_{(j)})$ in row i and column j is positive semidefinite whenever the column vectors $v_{(1)}, \ldots, v_{(m)}$ have the same dimension. This gives a device for constructing new kernels from an existing kernel. (Hint: See Example 11.6.1.)

Appendix: The Normal Distribution

A.1 Univariate Normal Random Variables

A random variable X is said to be standard normal if it possesses the density function

$$\psi(x) = \frac{1}{\sqrt{2\pi}} e^{-\frac{x^2}{2}}.$$

To find the characteristic function $\hat{\psi}(s) = \mathrm{E}(e^{isX})$ of X, we derive and solve a differential equation. Differentiation under the integral sign and integration by parts together imply that

$$\begin{aligned}
\frac{d}{ds}\hat{\psi}(s) &= \frac{1}{\sqrt{2\pi}} \int_{-\infty}^{\infty} e^{isx} ixe^{-\frac{x^2}{2}}\,dx \\
&= -\frac{i}{\sqrt{2\pi}} \int_{-\infty}^{\infty} e^{isx} \frac{d}{dx} e^{-\frac{x^2}{2}}\,dx \\
&= \frac{-i}{\sqrt{2\pi}} e^{isx} e^{-\frac{x^2}{2}} \Big|_{-\infty}^{\infty} - \frac{s}{\sqrt{2\pi}} \int_{-\infty}^{\infty} e^{isx} e^{-\frac{x^2}{2}}\,dx \\
&= -s\hat{\psi}(s).
\end{aligned}$$

The unique solution to this differential equation with initial value $\hat{\psi}(0) = 1$ is $\hat{\psi}(s) = e^{-s^2/2}$. The differential equation also yields the moments

$$\mathrm{E}(X) = \frac{1}{i}\frac{d}{ds}\hat{\psi}(0) = 0$$

and

$$\begin{aligned} \mathrm{E}(X^2) &= \frac{1}{i^2}\frac{d^2}{ds^2}\hat{\psi}(0) \\ &= \frac{1}{i^2}\Big[-\hat{\psi}(s)+s^2\hat{\psi}(s)\Big]_{s=0} \\ &= 1. \end{aligned}$$

An affine transformation $Y = \sigma X + \mu$ of X is normally distributed with density

$$\frac{1}{\sigma}\psi\Big(\frac{y-\mu}{\sigma}\Big) = \frac{1}{\sqrt{2\pi}\sigma}e^{-\frac{(y-\mu)^2}{2\sigma^2}}.$$

Here we take $\sigma > 0$. The general identity $\mathrm{E}\Big[e^{is(\mu+\sigma X)}\Big] = e^{is\mu}E\Big[e^{i(\sigma s)X}\Big]$ permits us to write the characteristic function of Y as

$$e^{is\mu}\hat{\psi}(\sigma s) = e^{is\mu-\frac{\sigma^2 s^2}{2}}.$$

The mean and variance of Y are μ and σ^2.

One of the most useful properties of normally distributed random variables is that they are closed under the formation of independent linear combinations. Thus, if Y and Z are independent and normally distributed, then $aY + bZ$ is normally distributed for any choice of the constants a and b. To prove this result, it suffices to assume that Y and Z are standard normal. In view of the form of $\hat{\psi}(s)$, we then have

$$\begin{aligned} \mathrm{E}\Big[e^{is(aY+bZ)}\Big] &= \mathrm{E}\Big[e^{i(as)Y}\Big]\,\mathrm{E}\Big[e^{i(bs)Z}\Big] \\ &= \hat{\psi}\Big[(a^2+b^2)s\Big]. \end{aligned}$$

Thus, if we accept the fact that a distribution function is uniquely defined by its characteristic function, $aY + bZ$ is normally distributed with mean 0 and variance $a^2 + b^2$.

Doubtless the reader is also familiar with the central limit theorem. For the record, recall that if X_n is a sequence of i.i.d. random variables with common mean μ and common variance σ^2, then

$$\lim_{n\to\infty} \Pr\left[\frac{\sum_{j=1}^n (X_j-\mu)}{\sqrt{n\sigma^2}} \le x\right] = \frac{1}{\sqrt{2\pi}}\int_{-\infty}^x e^{-\frac{u^2}{2}}\,du.$$

Of course, there is a certain inevitability to the limit being standard normal; namely, if the X_n are standard normal to begin with, then the standardized sum $n^{-1/2}\sum_{j=1}^n X_j$ is also standard normal.

A.2 Multivariate Normal Random Vectors

We now extend the univariate normal distribution to the multivariate normal distribution. Among the many possible definitions, we adopt the one most widely used in stochastic simulation. Our point of departure will be random vectors with independent standard normal components. If such a random vector X has n components, then its density is

$$\prod_{j=1}^{n} \frac{1}{\sqrt{2\pi}} e^{-x_j^2/2} = \left(\frac{1}{2\pi}\right)^{n/2} e^{-x^* x/2}.$$

Because the standard normal distribution has mean 0, variance 1, and characteristic function $e^{-s^2/2}$, it follows that X has mean vector $\mathbf{0}$, variance matrix I, and characteristic function

$$\mathrm{E}(e^{is^* X}) = \prod_{j=1}^{n} e^{-s_j^2/2} = e^{-s^* s/2}.$$

We now define any affine transformation $Y = AX + \mu$ of X to be multivariate normal [107]. This definition has several practical consequences. First, it is clear that $\mathrm{E}(Y) = \mu$ and $\mathrm{Var}(Y) = A\,\mathrm{Var}(X)A^* = AA^* = \Omega$. Second, any affine transformation $BY+\nu = BAX+B\mu+\nu$ of Y is also multivariate normal. Third, any subvector of Y is multivariate normal. Fourth, the characteristic function of Y is

$$\mathrm{E}(e^{is^* Y}) = e^{is^*\mu}\,\mathrm{E}(e^{is^* AX}) = e^{is^*\mu - s^* AA^* s/2} = e^{is^*\mu - s^*\Omega s/2}.$$

This enumeration omits two more subtle issues. One is whether Y possesses a density. Observe that Y lives in an affine subspace of dimension equal to or less than the rank of A. Thus, if Y has m components, then $n \geq m$ must hold in order for Y to possess a density. A second issue is the existence and nature of the conditional density of a set of components of Y given the remaining components. We can clarify both of these issues by making canonical choices of X and A based on the classical QR decomposition of a matrix, which follows directly from the Gram-Schmidt orthogonalization procedure [18].

Assuming that $n \geq m$, we can write

$$A^* = Q\begin{pmatrix} R \\ \mathbf{0} \end{pmatrix},$$

where Q is an $n \times n$ orthogonal matrix and $R = L^*$ is an $m \times m$ upper-triangular matrix with nonnegative diagonal entries. (If $n = m$, we omit the zero matrix in the QR decomposition.) It follows that

$$AX = (L \quad \mathbf{0}^*)\,Q^* X = (L \quad \mathbf{0}^*)\,Z.$$

In view of the usual change-of-variables formula for probability densities and the facts that the orthogonal matrix Q^* preserves inner products and has determinant ± 1, the random vector Z has n independent standard normal components and serves as a substitute for X. Not only is this true, but we can dispense with the last $n - m$ components of Z because they are multiplied by the matrix $\mathbf{0}^*$. Thus, we can safely assume $n = m$ and calculate the density of $Y = LZ + \mu$ when L is invertible. In this situation, $\Omega = LL^*$ is termed the Cholesky decomposition, and the usual change-of-variables formula shows that Y has density

$$\begin{aligned} f(y) &= \Big(\frac{1}{2\pi}\Big)^{n/2} |\det L^{-1}| e^{-(y-\mu)^*(L^{-1})^* L^{-1}(y-\mu)/2} \\ &= \Big(\frac{1}{2\pi}\Big)^{n/2} |\det \Omega|^{-1/2} e^{-(y-\mu)^*\Omega^{-1}(y-\mu)/2}, \end{aligned}$$

where $\Omega = LL^*$ is the variance matrix of Y.

To address the issue of conditional densities, consider the compatibly partitioned vectors $Y = (Y_1, Y_2)^*$, $Z = (Z_1, Z_2)^*$, and $\mu = (\mu_1, \mu_2)^*$ and matrices

$$L = \begin{pmatrix} L_{11} & \mathbf{0} \\ L_{21} & L_{22} \end{pmatrix} \qquad \Omega = \begin{pmatrix} \Omega_{11} & \Omega_{12} \\ \Omega_{21} & \Omega_{22} \end{pmatrix}.$$

Now suppose that Y_1 has full rank, and fix its value at y_1. The equation $y_1 = L_{11}X_1 + \mu_1$ shows that X_1 is fixed at the value $x_1 = L_{11}^{-1}(y_1 - \mu_1)$. Because no restrictions apply to X_2, we have

$$Y_2 = L_{22}X_2 + L_{21}L_{11}^{-1}(y_1 - \mu_1) + \mu_2.$$

Thus, Y_2 has conditional mean $L_{21}L_{11}^{-1}(y_1 - \mu_1) + \mu_2$ and conditional variance $L_{22}L_{22}^*$. To express these in terms of the components of $\Omega = LL^*$, observe that

$$\begin{aligned} \Omega_{11} &= L_{11}L_{11}^* \\ \Omega_{21} &= L_{21}L_{11}^* \\ \Omega_{22} &= L_{21}L_{21}^* + L_{22}L_{22}^*. \end{aligned}$$

The first two of these equations imply that $L_{21}L_{11}^{-1} = \Omega_{21}\Omega_{11}^{-1}$. The last equation then gives

$$\begin{aligned} L_{22}L_{22}^* &= \Omega_{22} - L_{21}L_{21}^* \\ &= \Omega_{22} - \Omega_{21}\Omega_{11}^{-1}L_{11}L_{11}^*\Omega_{11}^{-1}\Omega_{12} \\ &= \Omega_{22} - \Omega_{21}\Omega_{11}^{-1}\Omega_{12}. \end{aligned}$$

None of these calculations requires that Y_2 be of full rank. In summary, the conditional distribution of Y_2 given Y_1 is normal with mean and variance

$$\begin{aligned} \mathrm{E}(Y_2 \mid Y_1) &= \Omega_{21}\Omega_{11}^{-1}(Y_1 - \mu_1) + \mu_2 \\ \mathrm{Var}(Y_2 \mid Y_1) &= \Omega_{22} - \Omega_{21}\Omega_{11}^{-1}\Omega_{12}. \end{aligned} \tag{A.1}$$

References

[1] Acosta E, Delgado C (1994) Fréchet versus Carathéodory. *Amer Math Monthly* 101:332–338

[2] Acton FS (1990) *Numerical Methods That Work.* Mathematical Assn of Amer, Washington, DC

[3] Bartle RG (1996) Return to the Riemann integral. *Amer Math Monthly* 103:625–632

[4] Baum LE (1972) An inequality and associated maximization technique in statistical estimation for probabilistic functions of Markov processes. *Inequalities* 3:1–8

[5] Beltrami EJ (1970) *An Algorithmic Approach to Nonlinear Analysis and Optimization.* Academic Press, New York

[6] Böhning D, Lindsay BG (1988) Monotonicity of quadratic approximation algorithms. *Ann Instit Stat Math* 40:641–663

[7] Borwein JM, Lewis AS (2000) *Convex Analysis and Nonlinear Optimization: Theory and Examples.* Springer-Verlag, New York

[8] Botsko MW, Gosser RA (1985) On the differentiability of functions of several variables. *Amer Math Monthly* 92:663–665

[9] Boyle JP, Dykstra RL (1985) A method for finding projections onto the intersection of convex sets in Hilbert space. in *Advances in Order*

Restricted Statistical Inference, Lecture Notes in Statistics, Springer-Verlag, New York, 28–47

[10] Bradley EL (1973) The equivalence of maximum likelihood and weighted least squares estimates in the exponential family. *J Amer Stat Assoc* 68:199–200

[11] Bradley RA, Terry ME (1952), Rank analysis of incomplete block designs. *Biometrika*, 39:324–345

[12] Bregman LM (1965) The method of successive projection for finding a common point of convex sets. *Soviet Math Doklady* 6:688–692

[13] Bridger M, Stolzenberg G (1999) Uniform calculus and the law of bounded change. *Amer Math Monthly* 106:628–635

[14] Byrd RH, Nocedal J (1989) A tool for the analysis of quasi-Newton methods with application to unconstrained minimization. *SIAM J Numer Anal* 26:727–739

[15] Carathéodory C (1954) *Theory of Functions of a Complex Variable*, Vol 1. Chelsea, New York

[16] Censor Y, Zenios SA (1992) Proximal minimization with D-functions. *J Optimization Theory Appl* 73:451–464

[17] Charnes A, Frome EL, Yu PL (1976) The equivalence of generalized least squares and maximum likelihood in the exponential family. *J Amer Stat Assoc* 71:169–171

[18] Ciarlet PG (1989) *Introduction to Numerical Linear Algebra and Optimization*. Cambridge University Press, Cambridge

[19] Clarke CA, Price Evans DA, McConnell RB, Sheppard PM (1959) Secretion of blood group antigens and peptic ulcers. *Brit Med J* 1:603–607

[20] Conn AR, Gould NIM, Toint PL (1991) Convergence of quasi-Newton matrices generated by the symmetric rank one update. *Math Prog* 50:177–195

[21] Cox DR (1970) *Analysis of Binary Data*. Methuen, London

[22] Davidon WC (1959) Variable metric methods for minimization. *AEC Research and Development Report ANL–5990*, Argonne National Laboratory, USA

[23] Debreu G (1952) Definite and semidefinite quadratic forms. *Econometrica* 20:295–300

[24] de Leeuw J (1994) Block relaxation algorithms in statistics. in *Information Systems and Data Analysis*, edited by Bock HH, Lenski W, Richter MM, Springer-Verlag, New York, pp 308–325

[25] Dempster AP, Laird NM, Rubin DB (1977) Maximum likelihood from incomplete data via the EM algorithm (with discussion). *J Roy Stat Soc B* 39:1–38

[26] Dennis JE Jr, Schnabel RB (1983) *Numerical Methods for Unconstrained Optimization and Nonlinear Equations.* Prentice-Hall, Englewood Cliffs, NJ

[27] De Pierro AR (1993) On the relation between the ISRA and EM algorithm for positron emission tomography. *IEEE Trans Med Imaging* 12:328–333

[28] DePree JD, Swartz CW (1988) *Introduction to Real Analysis.* Wiley, New York

[29] de Souza PN, Silva J-N (2001) *Berkeley Problems in Mathematics*, 2nd ed. Springer-Verlag, New York

[30] Deutsch F (2001) *Best Approximation in Inner Product Spaces.* Springer-Verlag, New York

[31] Devijver PA (1985) Baum's forward-backward algorithm revisited. *Pattern Recognition Letters* 3:369–373

[32] Dobson AJ (1990) *An Introduction to Generalized Linear Models.* Chapman & Hall, London

[33] Duan J-C, Simonato J-G (1993) Multiplicity of solutions in maximum likelihood factor analysis. *J Stat Computation Simulation* 47:37–47

[34] Durbin R, Eddy S, Krogh A, Mitchison G (1998) *Biological Sequence Analysis: Probabilistic Models of Proteins and Nucleic Acids.* Cambridge University Press, Cambridge

[35] Dykstra RL (1983) An algorithm for restricted least squares estimation. *JASA* 78:837–842

[36] Edwards, CH Jr (1973) *Advanced Calculus of Several Variables.* Academic Press, New York

[37] Ekeland I (1974) On the variational principle. *J Math Anal Appl* 47:324–353

[38] Fang S-C, Puthenpura S (1993) *Linear Optimization and Extensions: Theory and Algorithms.* Prentice-Hall, Englewood Cliffs, NJ

[39] Feller W (1971) *An Introduction to Probability Theory and its Applications, Vol 2*, 2nd ed. Wiley, New York

[40] Fessler JA, Clinthorne NH, Rogers WL (1993) On complete-data spaces for PET reconstruction algorithms. *IEEE Trans Nuclear Sci* 40:1055–1061

[41] Fiacco AV, McCormick GP (1968) *Nonlinear Programming: Sequential Unconstrained Minimization Techniques.* Wiley, New York

[42] Fletcher R (2000) *Practical Methods of Optimization*, 2nd ed. Wiley, New York

[43] Fletcher R, Powell MJD (1963) A rapidly convergent descent method for minimization. *Comput J* 6:163–168

[44] Fletcher R, Reeves CM (1964) Function minimization by conjugate gradients. *Comput J* 7:149–154

[45] Flury B, Zoppè A (2000) Exercises in EM. *Amer Statistician* 54:207–209

[46] Forsgren A, Gill PE, Wright MH (2002) Interior point methods for nonlinear optimization. *SIAM Review* 44:523–597

[47] Geman S, McClure D (1985) Bayesian image analysis: An application to single photon emission tomography. *Proc Stat Comput Sec*, Amer Stat Assoc, Washington, DC, pp 12–18

[48] Gill PE, Murray W, Wright MH (1981) *Numerical Linear Algebra and Optimization, Vol 1.* Addison-Wesley, Reading, MA

[49] Golub GH, Van Loan CF (1996) *Matrix Computations*, 3rd ed. Johns Hopkins University Press, Baltimore, MD

[50] Gordon RA (1998) The use of tagged partitions in elementary real analysis. *Amer Math Monthly* 105:107–117

[51] Green PJ (1984) Iteratively reweighted least squares for maximum likelihood estimation and some robust and resistant alternatives (with discussion). *J Roy Stat Soc B* 46:149–192

[52] Green P (1990) Bayesian reconstruction for emission tomography data using a modified EM algorithm. *IEEE Trans Med Imaging* 9:84–94

[53] Grimmett GR, Stirzaker DR (1992) *Probability and Random Processes*, 2nd ed. Oxford University Press, Oxford

[54] Guillemin V, Pollack A (1974) *Differential Topology.* Prentice-Hall, Englewood Cliffs, NJ

[55] Hämmerlin G, Hoffmann K-H (1991) *Numerical Mathematics.* Springer-Verlag, New York

[56] Hastie T, Tibshirani R, Friedman J (2001) *The Elements of Statistical Learning: Data Mining, Inference, and Prediction.* Springer-Verlag, New York

[57] Heiser WJ (1995) Convergent computing by iterative majorization: theory and applications in multidimensional data analysis. in *Recent Advances in Descriptive Multivariate Analysis*, edited by Krzanowski WJ, Clarendon Press, Oxford pp 157–189

[58] Henrici P (1982) *Essentials of Numerical Analysis with Pocket Calculator Demonstrations.* Wiley, New York

[59] Herman GT (1980) *Image Reconstruction from Projections: The Fundamentals of Computerized Tomography.* Springer-Verlag, New York

[60] Hestenes MR, Stiefel E (1952) Methods of conjugate gradients for solving linear systems. *J Res Natl Bureau Standards* 29:409–439

[61] Hestenes MR (1981) *Optimization Theory: The Finite Dimensional Case.* Robert E Krieger Publishing, Huntington, NY

[62] Hiriart-Urruty J-B (1986) When is a point x satisfying $\nabla f(x) = \mathbf{0}$ a global minimum of $f(x)$? *Amer Math Monthly* 93:556-558

[63] Hochstadt H (1986) *The Functions of Mathematical Physics.* Dover, New York

[64] Hoel PG, Port SC, Stone CJ (1971) *Introduction to Probability Theory.* Houghton Mifflin, Boston

[65] Hoffman K (1975) *Analysis in Euclidean Space.* Prentice-Hall, Englewood Cliffs, NJ

[66] Hoffman K, Kunze R (1971) *Linear Algebra*, 2nd ed. Prentice-Hall, Englewood Cliffs, NJ

[67] Householder AS (1975) *The Theory of Matrices in Numerical Analysis.* Dover, New York

[68] Horn RA, Johnson CR (1985) *Matrix Analysis.* Cambridge University Press, Cambridge

[69] Horn RA, Johnson CR (1991) *Topics in Matrix Analysis.* Cambridge University Press, Cambridge

[70] Hunter DR, Lange K (2004) A tutorial on MM algorithms. *Amer Statistician* 58:30–37

[71] Jennrich RI, Moore RH (1975) Maximum likelihood estimation by means of nonlinear least squares. *Proceedings of the Statistical Computing Section: Amer Stat Assoc* 57–65

[72] Karlin S, Taylor HM (1975) *A First Course in Stochastic Processes,* 2nd ed. Academic Press, New York

[73] Keener JP (1993), The Perron-Frobenius theorem and the ranking of football teams, *SIAM Review*, 35:80–93

[74] Kelley CT (1999) *Iterative Methods for Optimization.* SIAM, Philadelphia

[75] Khalfan HF, Byrd RH, Schnabel RB (1993) A theoretical and experimental study of the symmetric rank-one update. *SIAM J Optim* 3:1–24

[76] Kingman JFC (1993) *Poisson Processes.* Oxford University Press, Oxford

[77] Kuhn HW, Tucker AW (1951) Nonlinear programming. in *Proceedings of the Second Berkeley Symposium on Mathematical Statistics and Probability.* University of California Press, Berkeley

[78] Kuhn S (1991) The derivative á la Carathéodory. *Amer Math Monthly* 98:40–44

[79] Lang S (1971) *Linear Algebra*, 2nd ed. Addison-Wesley, Reading, MA

[80] Lange K (1994) An adaptive barrier method for convex programming. *Methods Applications Analysis* 1:392–402

[81] Lange K (1995) A gradient algorithm locally equivalent to the EM algorithm. *J Roy Stat Soc B* 57:425–437

[82] Lange K (1999) *Numerical Analysis for Statisticians.* Springer-Verlag, New York

[83] Lange K (2002) *Mathematical and Statistical Methods for Genetic Analysis,* 2nd ed. Springer-Verlag, New York

[84] Lange K, Carson R (1984) EM reconstruction algorithms for emission and transmission tomography. *J Computer Assist Tomography* 8:306–316

[85] Lange K, Fessler JA (1995) Globally convergent algorithms for maximum a posteriori transmission tomography. *IEEE Trans Image Processing* 4:1430–1438

[86] Lange K, Hunter D, Yang I (2000) Optimization transfer using surrogate objective functions (with discussion). *J Computational Graphical Stat* 9:1–59

[87] Lehmann EL (1986) *Testing Statistical Hypotheses*, 2nd ed. Wiley, New York

[88] Little RJA, Rubin DB (1987) *Statistical Analysis with Missing Data.* Wiley, New York

[89] Luenberger DG (1984) *Linear and Nonlinear Programming,* 2nd ed. Addison-Wesley, Reading, MA

[90] Magnus JR, Neudecker H (1988) *Matrix Differential Calculus with Applications in Statistics and Econometrics.* Wiley, New York

[91] Mangasarian OL, Fromovitz S (1967) The Fritz John necessary optimality conditions in the presence of equality and inequality constraints. *J Math Anal Appl* 17:37–47

[92] Marsden JE, Hoffman MJ (1993) *Elementary Classical Analysis,* 2nd ed. W H Freeman & Co, New York

[93] McLachlan GJ, Krishnan T (1997) *The EM Algorithm and Extensions.* Wiley, New York

[94] McLeod RM (1980) *The Generalized Riemann Integral.* Mathematical Association of America, Washington, DC

[95] McShane EJ (1973) The Lagrange multiplier rule. *Amer Math Monthly* 80:922–925

[96] Meyer RR (1976) Sufficient conditions for the convergence of monotonic mathematical programming algorithms. *J Computer System Sci* 12:108–121

[97] Miller KS (1987) *Some Eclectic Matrix Theory.* Robert E Krieger Publishing, Malabar, FL

[98] Moré JJ, Sorensen DC (1983) Computing a trust region step. *SIAM J Sci Stat Comput* 4:553–572

[99] Narayanan A (1991) Algorithm AS 266: maximum likelihood estimation of the parameters of the Dirichlet distribution. *Appl Stat* 40:365–374

[100] Nazareth L (1979) A relationship between the BFGS and conjugate gradient algorithms and its implications for new algorithms. *SIAM J Numer Anal* 16:794–800

[101] Nelder JA, Wedderburn RWM (1972) Generalized linear models. *J Roy Stat Soc A* 135:370–384

[102] Nocedal J (1991) Theory of algorithms for unconstrained optimization. *Acta Numerica 1991*: 199–242

[103] Ortega JM (1990) *Numerical Analysis: A Second Course.* Society for Industrial and Applied Mathematics, Philadelphia

[104] Peressini AL, Sullivan FE, Uhl JJ Jr (1988) *The Mathematics of Nonlinear Programming.* Springer-Verlag, New York

[105] Press WH, Teukolsky SA, Vetterling WT, Flannery BP (1992) *Numerical Recipes in Fortran: The Art of Scientific Computing*, 2nd ed. Cambridge University Press, Cambridge

[106] Rabiner L (1989) A tutorial on hidden Markov models and selected applications in speech recognition. *Proc IEEE* 77:257–285

[107] Rao CR (1973) *Linear Statistical Inference and its Applications*, 2nd ed. Wiley, New York

[108] Royden HL (1988) *Real Analysis*, 3rd ed. Macmillan, London

[109] Rudin, W (1979) *Principles of Mathematical Analysis,* 3rd ed. McGraw-Hill, New York

[110] Schölkopf B, Smola AJ (2002) *Learning with Kernels: Support Vector Machines, Regularization, Optimization, and Beyond.* MIT Press, Cambridge, MA

[111] Smith CAB (1957) Counting methods in genetical statistics. *Ann Hum Genet* 21:254–276

[112] Sorensen DC (1997) Minimization of a large-scale quadratic function subject to spherical constraints. *SIAM J Optim* 7:141–161

[113] Strang G (1986) *Introduction to Applied Mathematics.* Wellesley-Cambridge Press, Wellesley, MA

[114] Swartz C, Thomson BS (1988) More on the fundamental theorem of calculus. *Amer Math Monthly* 95:644–648

[115] Tanner MA (1993) *Tools for Statistical Inference: Methods for the Exploration of Posterior Distributions and Likelihood Functions*, 2nd ed. Springer-Verlag, New York

[116] Thompson HB (1989) Taylor's theorem using the generalized Riemann integral. *Amer Math Monthly* 96:346–350

[117] Tikhomirov VM (1990) *Stories about Maxima and Minima.* American Mathematical Society, Providence, RI

[118] Titterington DM, Smith AFM, Makov UE (1985) *Statistical Analysis of Finite Mixture Distributions.* Wiley, New York

[119] Vapnik V (1996) *The Nature of Statistical Learning Theory.* Springer-Verlag, New York

[120] Weeks DE, Lange K (1989) Trials, tribulations, and triumphs of the EM algorithm in pedigree analysis. *IMA J Math Appl Med Biol* 6:209–232

[121] Whyte BM, Gold J, Dobson AJ, Cooper DA (1987) Epidemiology of acquired immunodeficiency syndrome in Australia. *Med J Aust* 147:65–69

[122] Wu CF (1983) On the convergence properties of the EM algorithm. *Ann Stat* 11:95–103

[123] Yee PL, Vyborný R (2000) *The Integral: An Easy Approach after Kurzweil and Henstock.* Cambridge University Press, Cambridge

Index

Springer Texts in Statistics *(continued from page ii)*

Lehmann and Casella: Theory of Point Estimation, Second Edition
Lindman: Analysis of Variance in Experimental Design
Lindsey: Applying Generalized Linear Models
Madansky: Prescriptions for Working Statisticians
McPherson: Applying and Interpreting Statistics: A Comprehensive Guide, Second Edition
Mueller: Basic Principles of Structural Equation Modeling: An Introduction to LISREL and EQS
Nguyen and Rogers: Fundamentals of Mathematical Statistics: Volume I: Probability for Statistics
Nguyen and Rogers: Fundamentals of Mathematical Statistics: Volume II: Statistical Inference
Noether: Introduction to Statistics: The Nonparametric Way
Nolan and Speed: Stat Labs: Mathematical Statistics Through Applications
Peters: Counting for Something: Statistical Principles and Personalities
Pfeiffer: Probability for Applications
Pitman: Probability
Rawlings, Pantula and Dickey: Applied Regression Analysis
Robert: The Bayesian Choice: From Decision-Theoretic Foundations to Computational Implementation, Second Edition
Robert and Casella: Monte Carlo Statistical Methods
Rose and Smith: Mathematical Statistics with *Mathematica*
Ruppert: Statistics and Finance: An Introduction
Santner and Duffy: The Statistical Analysis of Discrete Data
Saville and Wood: Statistical Methods: The Geometric Approach
Sen and Srivastava: Regression Analysis: Theory, Methods, and Applications
Shao: Mathematical Statistics, Second Edition
Shorack: Probability for Statisticians
Shumway and Stoffer: Time Series Analysis and Its Applications
Simonoff: Analyzing Categorical Data
Terrell: Mathematical Statistics: A Unified Introduction
Timm: Applied Multivariate Analysis
Toutenburg: Statistical Analysis of Designed Experiments, Second Edition
Wasserman: All of Statistics: A Concise Course in Statistical Inference
Whittle: Probability via Expectation, Fourth Edition
Zacks: Introduction to Reliability Analysis: Probability Models and Statistical Methods